Kieran Finnane is a founding journalist of the *Alice Springs News*, established in 1994, now publishing online. She contributes arts writing and journalism to national publications, including *Griffith Review, The Saturday Paper, Artlink* and *Art Monthly Australasia*. Her first book, *Trouble: On Trial in Central Australia*, was published in 2016.

PRAISE FOR *PEACE CRIMES*

'*Peace Crimes* is an engrossing and illuminating portrayal of six activists, a political trial and the secret machinations of the Five Eyes on Australian soil. Compelling to the core, Finnane is an excellent journalist who asks far more of the facts than mere repudiation; rather she asks for pause and reflection. I, for one, will never drive past or fly over these sites without thinking of all the unknowns, ever again.'

— ANNA KRIEN, AUTHOR OF *NIGHT GAMES* AND *ACT OF GRACE*

PEACE CRIMES

KIERAN FINNANE

PINE GAP, NATIONAL SECURITY AND DISSENT

First published 2020 by University of Queensland Press
PO Box 6042, St Lucia, Queensland 4067 Australia

uqp.com.au
reception@uqp.uq.edu.au

Cover design by Josh Durham (Design by Committee)
Cover photograph: 'Pine Gap (A photograph of the Centre of Australia)' by Kristian Laemmle-Ruff
Author photo by Erwin Chlanda
Typeset in 11.5/15 pt Bembo by Post Pre-press Group, Brisbane
Printed in Australia by McPherson's Printing Group

The University of Queensland Press is assisted by the Australian Government through the Australia Council, its arts funding and advisory body.

A catalogue record for this book is available from the National Library of Australia

ISBN 978 0 7022 6044 5 (pbk)
ISBN 978 0 7022 6222 7 (epdf)
ISBN 978 0 7022 6223 4 (epub)
ISBN 978 0 7022 6224 1 (kindle)

University of Queensland Press uses papers that are natural, renewable and recyclable products made from wood grown in well-managed forests and other controlled sources. The logging and manufacturing processes conform to the environmental regulations of the country of origin.

CONTENTS

For my parents

Patricia and Peter Finnane

PROLOGUE

Each night, here in the centre of Australia, I go outside to look at the sky, at its crystalline beauty, its movements and seasonal changes. A few years back, I learned about the Emu that comes out at night, taking shape from long dark matter in the Milky Way where stars are not. Wanta Steve Patrick Jampijinpa was talking to artists, art workers and art historians about Warlpiri ways of seeing. Warlpiri lands, including his home community of Lajamanu, lie to the northwest of Alice Springs, which is in the heart of Arrernte country. Using a digitally enhanced photograph he showed us the Southern Cross sitting on the Emu's head like a crown: 'The Queen's crown is really a late crown … there was another crown here,' he said. 'Stop being Australian, start being Australia. Be part of it, it's your birthright.'

The Emu is not always visible, but once seen, I can't unsee it.

On the night of 28 September 2016 I was at home, a bush block on the south side of Alice. While I was looking up at the sky, a group of five people were walking by the same starlight through shadowy bushland unfamiliar to them, some fifteen kilometres to the west. I would come to know them as the Peace Pilgrims – father and son Jim and Franz Dowling, Margaret Pestorius, Andy Paine and Timothy Webb.

Of other happenings in the world on that date, as countless as the stars, I single out one. In a village named Shadal Bazar, in the eastern

Achin district of Afghanistan, a man called Haji Rais was coming to grips with the deaths in his home of at least fifteen people. They had come to welcome him back from a pilgrimage to Mecca. After a meal of grilled sheep and a night of talk, everyone slept. The missile from a US drone struck before dawn. The survivors woke in a rain of shrapnel.

The strike was one of many aimed at Islamic State (Daesh) militants, whether by plane or drone, authorised that year by US president Barack Obama. Immediately, claims swirled that most of the dead and wounded were civilians. Survivors were taken to hospital in the provincial capital Jalalabad, where they spoke to reporters: 'I saw dead and wounded bodies everywhere,' Raghon Shinwari told them.

US Forces in Afghanistan (USFOR-A) confirmed a 'counter-terrorism' air strike, intended 'to degrade, disrupt, and destroy Daesh'. Its civilian toll was condemned by the United Nations: Daesh personnel might have died, but so had students and a teacher as well as members of families considered to be pro-government.

Civilian deaths raised questions of compliance with international legal obligations to protect them. By what criteria had the targets been identified? Had everything possible been done to avoid civilian casualties? Was the toll proportionate to the military objective of the attack? USFOR-A would say only that 'this was not a civilian casualty incident'.

Daesh had gained significant control over the Achin district, but the 'dozens of men' visiting survivors at the hospital insisted that the group was not operating within a mile of the village. 'If we were Daesh, do you think we would get together here?' asked Obaidullah, a university student. Mohabad Khan's lower body had been paralysed when shrapnel hit his spine. 'There is no Daesh in the village and every night the police go on patrols,' he said.

Haji Rais, owner of the targeted house, said nineteen people were injured in the strike; fifteen were killed. He later provided journalists with the names of the dead, matched their names with the names of their fathers and the districts they came from, writing his own name alongside the school principal's – Hakmatullah, his son.

~

This event, calamitous for those Afghans, would never have penetrated my consciousness if not for the Peace Pilgrims. What links us – me about to go to my warm bed, the Pilgrims walking through the late September night, and these people so far away, the dead and the survivors of a US drone strike, whether militant or civilian? The answer is what this book is about and its lynchpin is Pine Gap, the secretive US military base some nineteen kilometres southwest of Alice Springs. It was where the Pilgrims were heading.

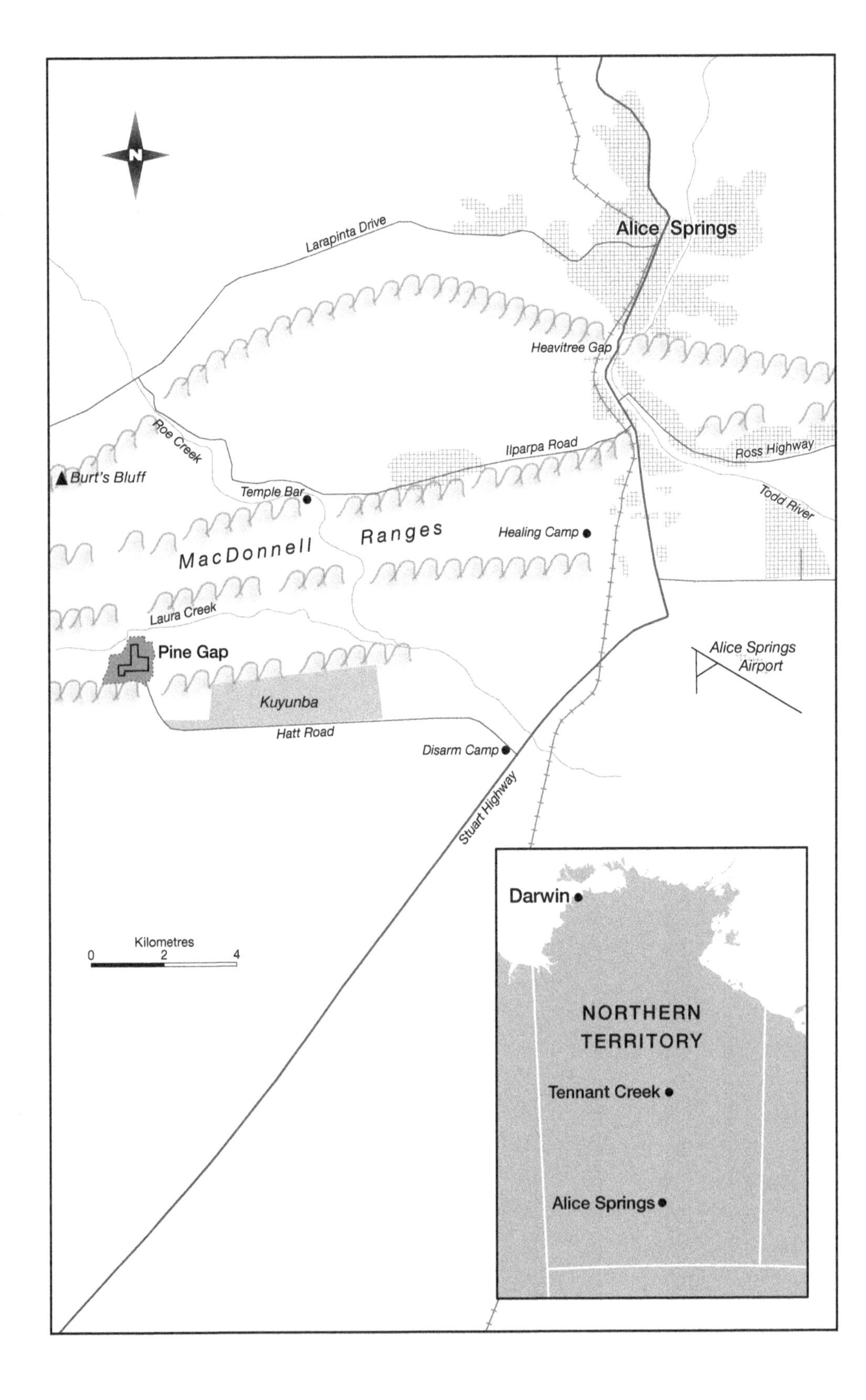
N
Alice Springs
Larapinta Drive
Heavitree Gap
Roe Creek
Ilparpa Road
Ross Highway
Burt's Bluff
Temple Bar
Todd River
Healing Camp
MacDonnell Ranges
Laura Creek
Pine Gap
Alice Springs Airport
Kuyunba
Hatt Road
Disarm Camp
Stuart Highway
Kilometres
0
2
4
Darwin
NORTHERN TERRITORY
Tennant Creek
Alice Springs

SEEING, NOT SEEING

The base had been brought into sharp relief for me in early 2016 by the work of a young artist from Melbourne, Kristian Laemmle-Ruff, titled *Pine Gap (a photograph of the Centre of Australia)*, 2015. To take it, Kristian had walked through the night, just as the Pilgrims would later do. It's clear that he must have gone through the outer fence and some way onto the base to reach his position, risking detection and prosecution. In making the image public, he took a further risk, exposing what he had obviously done in this prohibited area.

The exceptional digital clarity of his image reveals the military base in all its gleaming impersonality just as dawn arrives in this secluded valley. The sun's light seems to be waking in the ground itself while the artificial lights of the facility are still blazing, the image compressing into a single moment two emblematic orders of the centre of Australia, the natural and the man-made. The engineered dazzle of the military installation exudes an aura of control and power, man over nature, man over any opposing force.

Kristian made the photograph the signature work of his exhibition *Mind the Gap*, which showed in a number of galleries around Australia. He asked me to talk at the show's opening in Alice Springs. The title was an obvious reference to Pine Gap, but pointed also to the gaps in our understanding of the way the military base might be connected to such seemingly disparate matters as the value of local real estate, the tragic fallout of the Fukushima nuclear meltdowns of

March 2011, and the drying of ancient mound springs and wetlands in remote South Australia. These were among the subjects spanned in other photographs included in the show. The link between them, briefly, is nuclear power – civil and military – and the way it has altered what we value in land as well as altering the land itself.

The artist has lineage in the peace movement, but when he talks about his work he goes first to his own developing consciousness and perceptions and where they are taking him. Not that he hesitates, when I ask about how his focus on militarism took root, to talk about his father. Tilman Ruff AM is a physician and professor. In 2006, he co-founded the International Campaign to Abolish Nuclear Weapons, known by its clever acronym, ICAN. In July 2017, in partnership with a group of progressive countries, ICAN succeeded in getting the United Nations to adopt a treaty banning nuclear weapons. Support was 122 votes to one; missing from the vote were the nuclear powers and their supporters including Australia, which did its best to derail the process. Later that year, ICAN were awarded the Nobel Peace Prize for their efforts leading to the treaty and for drawing attention more broadly to the catastrophic humanitarian consequences of any use of nuclear weapons.

Not surprisingly then, Tilman Ruff's son had grown up more than usually aware of militarism and its impacts. And he'd long heard talk of Pine Gap. In 2014, aged in his mid-twenties, Kristian set out on a hitchhiking trip around Australia, a 'coming of age' journey. He'd been working as a commercial photographer since he was nineteen and doing 'passion projects' on the side – art photography that he exhibited in group and solo shows – and now he wanted to get a more direct personal understanding of the country he'd grown up in. Through his father he'd had contact with Richard Tanter, a senior academic political scientist and board member of ICAN. At the time, Richard, together with researchers Desmond Ball and Bill Robinson, was writing a report on the activities of Pine Gap, the seventh in a series of eight for the California-based Nautilus Institute. Richard told Kristian about his own experience of photographing the base from a peak west of town called Burt's Bluff. It had been important for Richard, not just for the images he got but for the completely different sense of the topography it gave him. It was 'a kind of a ground-truthing'.

Richard's photographs, though not great technically, had provided solid evidence and usable images of the exact character of the Torus quasi-parabolic multi-beam antenna that he had detected a year or so earlier through Google Earth. He and his colleagues have done more poring over Google Earth and other images from the web than they care to think about. Not long ago, he spotted a new high-frequency mast at Pine Gap. Actually, it was likely to have been there for eight or nine years but he hadn't noticed it before. He was keen to talk to Bill Robinson about it, but then 'this stuff changes all the time' and 'antennas aren't really important, they're the mushrooms for the fungi, the real action is in the fungi, but the mushroom allows you to see it'.

After identifying the Torus at Pine Gap and armed with information from files leaked by former National Security Agency employee Edward Snowden, Richard and co. soon identified other Torus antennas, installed at US bases around the globe: it was proof of an increasingly comprehensive Five Eyes surveillance system of which Pine Gap is a part. Its Torus made for continuity in purview with the installations in Seeb, Oman, and Waihopai, New Zealand; the rest of the world was surveilled from Bude and Menwith Hill in the UK. The grainy long-lens image of the Torus that Richard published showed its unmistakeable half-pipe form behind a cluster of three radome-covered antennas, against the equally unmistakeable rust-coloured earth of the Centre.

Looking at Richard's photograph, Kristian thought he could do better, in part down to having a much better camera and lenses. He borrowed a friend's bicycle and headed west out of Alice along Larapinta Drive. The heat of the November day was easing by the time he knocked on the door of the landowner and asked permission to cross the flat and climb the bluff. The man who answered didn't exactly give permission but nor did he emphatically deny it. Kristian decided to keep going. He hid the bike in some bushes and hiked up into the hills. When he got to the crest of the bluff he realised that a photograph from there would not offer anything new. A lot of the ground-level features of the installation were still obscured. He needed to get closer. The moon would be coming up after midnight. With its light he wouldn't need a torch so he lay down and tried to get some sleep.

Around 2 am he started down Burt's Bluff. With a heavy pack on his back, loose rock underfoot and only the moon lighting his way, the descent was tricky. He crossed another flat and then clambered up the last rise ahead of the inner boundary of the base: laid out before him was the unimpeded view he wanted. He kept his head down and watched. A patrol vehicle did constant laps of the perimeter, with a top-mounted searchlight beaming into the darkness. His window of opportunity would be very small. He waited for dawn to come. As it started to light up the landscape he took his images and ran.

It was only afterwards that the potential legal ramifications sank in. Lawyers told him the images were 'high risk' and advised against publishing. When he got back to Melbourne he showed them to Richard. Their clarity and precision of all parts of the base would help Richard and his colleagues confirm their analysis to date. Richard also found them 'diabolically beautiful': 'I was sure that once they were out they would become, in that over-used word, iconic – as they justly have.'

~

There were three radomes at the base when operations began in 1970, housing antennas that collected electronic signals intelligence downlinked from satellites. The domes have become Pine Gap's most recognisable feature but really they are of the least concern. Their main purpose is to shield the antennas from dust and weather, although they also prevent snooping eyes from seeing where the antennas are pointed. From the start the antennas had both defensive and offensive roles relating to US nuclear war-planning. In the early Cold War years the central focus was on finding out about Soviet ballistic missile testing sites, launch bases and radars. In the event of a US offensive, the signals information would also have helped their nuclear-armed B-52s to avoid or block Soviet radar detection. Later, when arms control agreements with the USSR were reached, the signals intel assisted with their verification. (This aspect of Pine Gap's mission is the one the Australian Government, of whatever stripe, likes to emphasise.)

Richard uses the image of big eyes and big ears to explain the technology of the base. Big infra-red eyes are the key to its role in nuclear-fighting: telescopes on satellites that look for the heat bloom of intercontinental ballistic missiles (ICBMs), to give the US early warning – just thirty minutes – of an imminent attack. The first ground station in Australia for monitoring ICBMs was at Nurrungar in South Australia. When it was closed in 1999, the ground station was moved to Pine Gap. By then, US control of the base's operations had passed from the CIA, an independent agency, to the Department of Defence's National Reconnaissance Office, in line with the increased militarisation of the base – in personnel and mission – and reflecting the shift in US strategic priorities towards wars of intervention and counterterrorism. New and upgraded satellites with powerful infra-red sensors joined the old, with the whole system known as overhead persistent infra-red or OPIR. It feeds data to US military command for early warning to US–Japan missile defence systems, and for targeting in the event of an American nuclear attack. The missiles fly at extraordinary speed through space, and the 'cueing' via Pine Gap is essential for them to have any chance of finding their targets. This function is actually operated remotely from Buckley Air Force Base in Colorado, with the data flowing between bases by optical fibre and satellite communication. The relay ground station for OPIR at Pine Gap is in a fenced-off area at the western edge of the base, accessed only by a very small number of staff, essentially for undertaking maintenance.

The big ears are three Advanced Orion satellites sitting above Indonesia and the Indian Ocean, collecting a huge volume of signals from the middle of the Pacific to the western edge of Africa. Turned up to them and to a host of communications satellites are, as of February 2016, nineteen radome-covered antennas, some much bigger than others, as well as fourteen uncovered systems. These include the powerful Torus multi-beam antenna that Richard had gone to such trouble to photograph. Installed in 2008, it is capable of monitoring thirty-five or more satellites at a time. The eavesdropping used to be more tightly focussed, but now Pine Gap, as a critical link in the Five Eyes system, is involved in automated mass surveillance of communications around the globe – of friends and foes, civilian and military.

Calls by cell phone and sat phone as well as emails are analysed and, where deemed relevant, passed on to military command or, outside zones of armed conflict, to the CIA. Targeted killing in such contexts is legal only if it is strictly and directly necessary to save life, yet 'signature' targeting has been a thing for the CIA, based on surveillance and analysis of behaviour ('pattern of life') rather than a precise identity. In zones of armed conflict, unless it can be established that the targeted person has a 'continuous combat function' or is 'directly participating in hostilities', an attack on them is unlawful. Only 'near certainty' of a target's status is required, a standard introduced by President Obama in response to controversy over the targeted killing program. Think of it relative to the standard of 'beyond reasonable doubt' underpinning capital punishment in the US. In countries that neither Australia nor the US is at war with, the strikes are arguably extrajudicial assassinations, in which Australia, through Pine Gap, is involved. As Richard puts it, Australia is 'literally and institutionally hard-wired' into both the surveillance system and the military operations, whether conventional or covert; there is no transparency around either.

For that drone strike on Shadal Bazar, I asked Richard, what is the likelihood that Pine Gap was involved in supplying the targeting data? Eighty per cent, fifty per cent, maybe not at all?

In the 1980s and '90s he could have said with one hundred per cent certainty that Pine Gap was involved, with a qualification that he would come back to. Then, Pine Gap was a so-called stand-alone system. It was connected to its three Orion series signals intelligence satellites; Menwith Hill in Yorkshire, UK, was connected to another three. They didn't talk. Pine Gap downlinked its three and did some processing, or shunted the data off to Fort Meade in the US.

What has happened since makes it much more difficult to answer with certainty. Now all of the ground stations are linked to each other: Pine Gap is linked to Menwith Hill and Government Communications Headquarters (GCHQ) in Britain; to Buckley, Colorado, and Fort Meade, Maryland, in the US; and to Waihopai in New Zealand. Oman and Cyprus are mainly reception stations, but the others all exchange tasking: 'Pine Gap will say, "We've got too much work – all this stuff, can you process it?" Or one station will be charged with downlinking, and the processing will happen elsewhere, or you will have situations where, pretty clearly in the Middle Eastern area, there is overlap in coverage between Pine Gap and Menwith Hill.'

The qualification Richard made is that the listening is done not only by satellites and this has been so for a long time: 'The Americans have aircraft flying around from Japan to North Korea all the time and there are electronic surveillance ships they can bring in for certain occasions. Pine Gap and Menwith Hill suck up vast amounts of data but they are not the only platforms.'

If there were transparency about who targeted what and how, I asked him, would it be retrievable information?

'Oh yes, absolutely, they know.'

With Pine Gap's involvement in the use of military force, including targeted killings by drone, there has never been a fully exposed smoking gun. Yet.

Richard believes a leak is inevitable: 'Sooner or later there'll be another Edward Snowden, stuff will come out, somebody will say, "I was on a desk at Pine Gap; I saw or took part in conversations with targeting authorities; location data and characterisation data that we produce at Pine Gap was communicated to drone operation authorities." It's very hard to say that's not illegal under Australian law.'

While avoiding specific denials, Australian politicians simply bat away questions about this. In face of the compelling evidence and arguments put into the public domain by academic researchers, journalists, lawyers and the rare political colleague, they refuse to comment. Together with generalised claims about Pine Gap's importance for Australia's security and the 'pervasive threat of terrorism', to quote a recent defence minister, this has been enough to mute public debate, nationally and in Alice Springs. Here the general attitude is best reflected in the Town Council's one sentence policy about the facility: 'Council supports the retention of the Australian/American Joint Defence Facility Pine Gap, and acknowledges the importance of this Facility for the defence of Australian territory and for the economic and social benefit of Alice Springs.' In relation to the defence of Australia, the council's policy is largely unchanged from when it was first formulated more than thirty years ago. A 2008 revision added the clause about 'economic and social benefit'. This has never been the subject of in-depth scrutiny.

Richard's research on the base is about letting in light on a largely unseen world: not only the materially existing, tightly guarded Five Eyes bases, and Pine Gap in particular, but the work they do, mostly invisible to the eye and enormously challenging to understand in its density and complexity. It needs images. This was driven home to him when he first saw the now famous photograph of Pine Gap's counterpart base at Menwith Hill, taken by geographer and artist Trevor Paglen and installed at massive scale in the Gloucester Road tube station in London. Spliced between the dirty red-brick arches along the platform was what appeared to be a lovely photograph of the English countryside – gentle green fields enclosed by low stone walls, stone houses, soft light. It took a little while before he registered, amazed, the faint radomes looming on the skyline: 'Paglen – as he so often does – had managed to say something important about the way we see/don't see these places he calls blank spots on the map.'

Kristian's images of Pine Gap are of the same order as Paglen's, unprecedented in their coverage, quality and power. They take their place within a certain canon: 'All of the publicly important work about Pine Gap has been rooted in imagery,' says Richard. 'There is almost no other way to convey some notion of what the place is about, absent patience to read a lot of words.'

The publication of Richard's own informational photograph of the Torus in May 2015 was in part a test of whether the Australian Government, then under Tony Abbott, would prosecute. When that didn't happen, the way felt much clearer for Kristian's photographs to be included in the new report for the Nautilus Institute, published online in early 2016. At the same time Kristian released his photographs under a Creative Commons licence, which made them freely available to the media. They were quickly republished by a number of major news outlets reporting on the Nautilus research. Kristian was pleased to have them in the world at last.

~

I had described Kristian's photograph of Pine Gap as 'arresting' before I came across John Berger using that word, as literally the most apt, for the effect on us of images of violence: 'They bring us up short.' Berger was writing, in 1972, about the kind of photographs that began to be published during the Vietnam War, that show what happens when people are blown up by bombs, photographs that show 'agony'. The despair that may engulf us when we look at them serves no point, Berger contended. The indignation that we might alternatively feel would require us to act, but as we return to our lives we realise the hopeless inadequacy of our possible actions, unless they are political. This would mean challenging 'the conduct of wars waged in our name'. To do this, it seems he saw acts of civil disobedience as essential, as there were no legal opportunities in 'the political systems as they exist'.

Kristian's photograph distils a refined violence, the cool corporatised face of ever more remote forms of warfare. Remote for the aggressor, that is; on the ground it still ends in bombs and blood. If we go there, imaginatively, when we look at Kristian's image, we have to ask ourselves what we think about that. But how does he get us to go there, to think?

Berger argued that there was work that photographers themselves could do to take their images beyond 'arrested moments' that have a linear message. They could 're-create context' for their image, so that it operated 'radially' like memory does, with many associations leading to and from it: 'A radial system has to be constructed around the photograph so that it may be seen in terms which are simultaneously personal, political, economic, dramatic, everyday, and historic.'

Such a system could include things like text – the title and artist's statement for Kristian's image are significant – but key contributions are made within the visual work itself. The photograph was taken as dawn arrived, a daily experience of the transitory, intimately familiar to us; as present in this place, which is generally prohibited from view, as it is in the places of our everyday lives. This acts against the aura of impenetrability of the installation, against the mystification of its power; it shares a common ground with us. In the art gallery, Kristian took this direction further. The photograph was presented in an immaculate light-box, emphasising the high-tech gloss of the base while adding another layer to the play of light in the scene. He made the frame for the light-box from natural wood, shifting the aesthetic between the natural and man-made, in keeping with the shifts contained within the image. He photoshopped into the foreground gleaming narrow bars that seem to reflect the dawn light, signalling that this was more than superbly executed visual reportage of a set of facts *out there*. The bars fittingly reference a fence but they also emphasise the position of the artist, an appeal to the viewer to stand alongside him and look at, as Kristian's artist's statement spelled out, 'a terror facility right in the middle of our country' – to either think about what goes on in that place or admit that we've pulled the blinds down.

These were the things on my mind as I prepared for opening Kristian's show. 'It can sometimes take a stranger within our midst to refocus our attention on that banal evil from which it never should have strayed.' That's how I began my speech. I was trying – with somewhat elevated diction! – to draw my audience into a moral as well as political reflection. For that was how I felt drawn. In three decades in Alice Springs I had done little serious reporting about Pine Gap and I was asking myself why. The nicknames 'space base' or 'spy base' served to obscure its military role in popular consciousness, mine too; its American personnel – unarmed, no uniforms – were part of the community. At the *Alice Springs News*, the weekly paper my partner Erwin Chlanda and I began in 1994, we occasionally employed family members of American base staff;

Americans were parents of some of our children's classmates, an American was our son's basketball coach. For other locals, Americans were their neighbours, members of their club or their church. Like many, I seemed to have adopted the local etiquette of 'not talking about the war'. We rarely embarrassed the Americans in our midst with probing questions about their jobs; the Australians who worked at the base in roughly equal numbers were mostly hidden in plain view.

Kristian's photograph stayed with me. I taped a small copy to my desk, a larger one to a wall in the kitchen. Sometime later, a friend put into my hands Sarah Sentilles's *Draw Your Weapons*, a long meditation on the power of photographs and what they require of us in response. In the context of the wars of this century, she revisits the thinking of Berger, Susan Sontag and more recent others who have asked themselves this question. Especially preoccupying for her is the invasion of Iraq, in which surveillance photos played such a critical role, and drone warfare, to which the camera is essential. Predator drones are equipped with three different types of camera; the images they produce are used to kill, but this is not new, photography has been used in this way since its invention, writes Sentilles: 'Teaching viewers how to see or not see each other. Teaching viewers how to treat one another.' That work does not all go in one direction, of course; in the space between seeing and not seeing is the creative possibility – for image-makers and viewers, for activists and writers too – of seeing differently, seeing anew. And of choosing not to kill.

Sentilles had called herself a pacifist most of her life, 'as if being against the wars my country fights means they have nothing to do with me'. Two photographs had jolted her from that position: one notorious, of an Iraqi man, bag over his head, wires attached to his body, tortured by American soldiers inside Abu Ghraib prison; the other of a joyful old man holding the violin he had started to make while interned as a conscientious objector during the Second World War. The violin-maker and other activists, artists and writers whom Sentilles encountered reminded her that 'The world is made. And can be unmade. Remade.' Kristian's photograph shows us a place that has been made, and it calls for an unmaking, at the very least of the idea that Pine Gap has nothing to do with me, with us.

~

Kristian staged his show in the lead-up to the 2016 annual conference of the Independent and Peaceful Australia Network (IPAN), held in Alice Springs. This network of community, faith and peace groups, trade unions and concerned individuals is opposed broadly to the presence of US forces on Australian soil, including the Marine 'rotations' in Darwin, first welcomed by Julia Gillard's Labor government as part of President Obama's 'pivot' to Asia. Campaigning against the Marines' presence remains an important part of IPAN's efforts. In 2016, though, it being the fiftieth anniversary of the Australian–US treaty that had established the base at Pine Gap, they turned their full attention to 'Secrets in the Centre'.

Richard Tanter gave the keynote address. He was blunt about Pine Gap's contemporary functions: 'Targeting data supplied by Pine Gap is used for drone killings both in war zones in Afghanistan and Iraq and in countries with which Australia is not at war, such as Pakistan, Somalia and Yemen. Not only does this make Australia culpable for well-documented war crimes, but it also generates cycles of terrorism in response.'

I had read and seen reports in mainstream media about drone warfare and was repelled by its asymmetric nature: operatives administering death by keystroke from an air-conditioned room in the US, with the targets and 'collateral damage' inevitably in non-Western countries. It was more troubling to think of the kill chain starting in an air-conditioned room so close to my home.

At that stage the last time that the military role of Pine Gap had received any vigorous local attention was thanks to other 'strangers in our midst' – a group calling themselves Christians Against All Terrorism. In December 2005 two of them succeeded in penetrating the base's operations area – 'the most serious security breach in Pine Gap's history', according to David Rosenberg, a former National Security Agency employee at the base. They sent photographs they took inside to eighteen Australian newspapers. The *Alice Springs News*, then still a print weekly, stood up to threats of prosecution and immediately published an image (most others did not). It showed 29-year-old Adele Goldie sitting on a roof beneath a dish-antenna, smiling and with her left hand raised in the peace sign. The shot was grainy and over-exposed in the foreground, but there was just enough definition to see the telltale forms of three radomes rising in the background, making the photograph worth its weight in activist gold. Jim Dowling, then fifty, had taken it. Erwin wrote, 'The flash from his camera could well have been from a rocket propelled grenade had the "insurgents" been from Al Qaeda instead of the Christians Against All Terrorism.' No doubt this was why it was 'The picture they didn't want you to see', his headline.

The threats against the *News* disappeared, but not so for the activists. For the first time the Commonwealth resorted to prosecuting protesters under the *Defence (Special Undertakings) Act 1952*. In fact the Act had never been used for any kind of prosecution before. The legal proceedings were protracted, with a maximum penalty of seven years in gaol hanging over the defendants' heads. They were convicted and fined, refused to pay and did gaol time in lieu, before finally winning an appeal against their principal convictions. For all their courage, however – and that of Edward Snowden, and the steady work of so many in civil society – military operations at the base not only continue but have expanded. Their high-tech nature should not obscure their real-world impacts. People die violent deaths as a result of the activities at Pine Gap.

'What counts as a liveable life and a grievable death?' asked philosopher Judith Butler in post-9/11 America. She was pondering the seeming unreality of non-Western peoples, particularly practitioners of Islam whose deaths by military means are so easily accepted – like the fifteen Afghans who died by drone strike in Shadal Bazar. I'd read their names, their relationships with one another, what happened; I'd looked at photographs. Still they and the whole situation felt unreal. I could have substituted, without effect, other names and places from their world, *not mine*.

'If violence is done against those who are unreal, then, from the perspective of violence, it fails to injure or negate those lives since those lives are already negated,' Butler wrote. 'But they have a strange way of remaining animated and so must be negated again (and again). They cannot be mourned because they are always already lost or, rather, never "were", and they must be killed, since they seem to live on, stubbornly, in this state of deadness. Violence renews itself in the face of the apparent inexhaustibility of its object.'

Walking not far from my home through that September 2016 night, the Peace Pilgrims keep it simple. They lament and pray for the dead, especially those who have died by drone strike. They are particularly repelled by this high-tech killing, guided by the signals intelligence collected by the network of which Pine Gap is a critical part. These are grievable deaths, the Pilgrims are intent on saying.

A hasty photograph taken of them under a streetlight, en route to the starting point of their walk, shows them warmly dressed in dark clothes, smiling, purposeful. Past midnight and into the next morning they walked, getting lost, fighting exhaustion. Around 4 am they finally reached the outer fence of Pine Gap. They stepped through, climbed the hill that loomed in front of them, praying as they went and playing a lament for the war dead, until they were arrested. They had damaged nothing, hurt no-one, but they'd done it on a 'prohibited area for defence purposes'. Like the Christians Against All Terrorism before them, they were charged under the *Defence (Special Undertakings) Act*.

A bit over one year later I was following their trial in the Supreme Court of the Northern Territory.

IN THE DARK

The Admitted Facts are as dry as desert dust: the 'accused persons' were found within the Joint Defence Facility Pine Gap prohibited area on 29 September 2016; the CCTV security footage, showing what happened from the time of detection to arrest, runs from 3.30 am to 5 am.

The live action began hours before that, late on the night of the 28th, a Wednesday. It was exhausting and exhilarating, fitting for a pilgrimage. They were answering a call to lament, witness and pray in the presence of this 'war-making place'.

Arrernte people know the area as Kweyernpe (Kuyunba on most maps). A conservation reserve by that name is at the southern boundary of the base, and within it is an important sacred site complex. Just beyond the reserve entrance, a sign for the base tells drivers to 'turn around now', 'official entry only', 'no photography from this point on'. There is no law invoked on this sign, although the *Defence (Special Undertakings) Act* might apply; it has provisions for people being 'in the neighbourhood' of a prohibited area. As far as I know there have been no consequences for people ignoring the sign's direction and going as far as the front gate about a kilometre further on, other than likely having their numberplate checked and possibly being questioned. This can happen even if you are simply visiting the Kweyernpe reserve. A surveillance camera is permanently mounted near the carpark and several people have reported being surveilled by drone as they walked through the reserve, with some of them also questioned by base security. None of this happened the day I visited with a friend.

We turned off Hatt Road into the reserve along a short dirt-road corridor, cattle fence either side, and suddenly we were in a different world. A long, narrow vale of soft sand between walls of coloured sandstone – cream, orange, rust, and in places darkened by lichen. Here and there the rockwalls form into sheltering overhangs and caves, fringed by the native pines that have given Pine Gap its name. The friend I was walking with picked up stone chips scattered across the sand and showed me where they had been worked as tools – the facets, the bulge forced by the blows of the hammer stone, the cutting edges. She pointed out the woollybutt grass whose seed was once an important food source. She spotted the remnants of grindstones which women used to make seed pastes.

The vale drew us in. Its rock art was faintly discernible. There were signs of long-ago logging and more recent campfires, but the place lately seemed seldom disturbed by people. More 'roo tail marks than human footprints, while thick carpets of leaves and small cones spread under the pine trees and piled up on the stone walls behind them. The vale took a bend. The late afternoon sun was throwing its long light onto the west-facing surfaces – the sandstone turned fiery, a pale-barked eucalyptus stood luminous like a sentinel just beyond the bend. My friend and I could go no further: a sign told us the sacred site was off limits to women and children, if we were not men to return to the carpark, and so we did.

It is a more than strange juxtaposition, these two areas of prohibition, both seats where secret male power is wielded – one ancient and now heavily circumscribed, the other, alien in the truest sense of the word and, via its extraordinary technology, global in its reach.

Protests about the military base, when they occur, are typically at the front gates, just this side of the ranges that shield the base infrastructure from view. The Peace Pilgrims, though, will approach from the north.

~

It's well after 10 pm when they start walking, leaving Larapinta Drive in the dark, without using their torches. The waning moon's narrow sliver won't appear till close to dawn but that means the stars are bright. The first stretch is through private land owned by a local who won't hesitate to call police if he becomes aware of their presence. He did call on their first attempt, not that the Pilgrims knew. That was three nights earlier, on Sunday. They had covered the distance more quickly than expected and, as they wanted to stage their action close to dawn, they lay down for a nap. Looking back, Margaret Pestorius laughs: 'Stay awake with me, Peter, John, James!' Her references tend to be musical or Christian or both. This one is to 'Tim Rice's Gospel', as she calls it, *Jesus Christ Superstar.*

The night was cold, they had no blankets, the ground was full of burrs and rocks but still they managed to fall asleep, like babes in the wood. It might have been an hour later that Margaret woke to hear vehicles approaching. She hissed to the others to keep their heads down. They laughed – too late! Police, federal and local, had already surrounded them and soon they were being hustled into a minibus and taken back to their camp. The next day they were served trespass notices, warning that if they went back onto that land, they would be prosecuted.

Along with scores of other peace activists, the Pilgrims had been drawn to Alice Springs by protest events marking the fiftieth anniversary of the US–Australian agreement that established the Pine Gap facility. The base would hold its celebrations the following year at the Alice Springs Convention Centre – strictly invitation only. Media were excluded despite the attendance of senior federal and territory politicians. It was a 'stealth party', quipped Erwin in the *Alice Springs News.* Hours before guests arrived, the Convention Centre was in lockdown, a large number of armed uniform and plain clothes police were inside and a drone was in the air – presumably for surveillance.

When the Close Pine Gap Convergence got underway the previous September, most visiting activists stayed in either of two camps set up en route to Pine Gap. From there they staged protest actions outside the base as well as in town itself. In the early hours of 29 September, at the same time as the Pilgrims were entering Pine Gap, in the Alice Springs industrial area four protesters chained themselves to the gates of US military contractor and weapons manufacturer Raytheon, under a banner reading 'Raytheon War Criminals'. Police minimised the profile of the action when no arrests were made. On Hatt Road a few days earlier, when a trio of Quaker grandmothers in their bonnets and quaint dress had staged a tea party, police had again taken a hands-off approach, directing traffic around them. Many protesters reported that police 'killed us with kindness', but as their actions unfolded organisers sent out real-time images to media, guaranteed to get some coverage in this era of the 24-hour news cycle.

As bold as these actions were, breaching the boundary of the base was a step further and, as the Pilgrims see it, not all are called to take it.

'Jesus called us,' says Jim Dowling. 'Blessed are the peacemakers.'

Jesus has been calling Jim to Pine Gap for a long time. Of course civil society has also had something to do with it, but he finds his main motivation in his faith. He believes utterly in the sanctity of human life, uniting his commitment to nonviolent activism for peace and against war with commitments across the spectrum of life, against poverty and abortion, to name two that loom large for him and his wife, Anne Rampa. They know their position on abortion isolates them from many in the peace movement, but they embrace it as a 'consistent life ethic', a moral position articulated by American Catholic bishops in the 1980s. The bishops saw that concern for the rights of the unborn could not stop at birth; their quality of life also had to be supported, if necessary by public programs (anathema for many conservatives in America). Conversely, life-denying policies and actions had to be opposed, meaning not only abortion but also the death penalty and nuclear war.

It was Pine Gap's contribution to 'nuclear preparedness' that first sparked Jim's opposition to its activities, which has only intensified with the base's role in the US drone program coming to light. This is what Jim hopes to at least disturb by his actions and he is not waiting for the church in Australia to lead. Frankly, he sees the church's silence on the issues as contributing to Australia's creeping militarism.

If Jim walked into a gathering of my family, especially the family on my father's side in the memory I carry of them, he'd seem to me right at home. It's the Irish look of him, the stamp on him of his strong Catholic belief, his large family, his embrace of voluntary poverty. When he explains to me why he goes barefoot – an act of solidarity with the poor who 'can't afford shoes' – it brings to mind my father's stories of growing up in the 1920s and '30s as one of nine children, going to school in bare feet and hand-me-down clothes; of his mother, whose only footwear was a pair of slippers. Poverty wasn't voluntary for Dad's family, and neither was it for Jim's. His father's experiences during the Second World War seem to have broken his mental health; he was never able to adequately support his wife and seven children. Jim recalls the embarrassment of going to school 'with the seat out of my pants', but schooling as well as post-war prosperity helped lift most of his siblings out of poverty. Only Jim has chosen the radical renunciation of material comforts that is part of the Catholic Worker ethos.

The Catholic Worker Movement began in America during the Great Depression, co-founded by Dorothy Day and Peter Maurin. Aiming to transform both individuals and society, it is about 'nonviolence, personal responsibility of all people to the poorest ones among us, and fidelity to community and to God'. Day was an ardent pacifist, even to the extent of opposing military involvement in the Second World War, though not all in the movement could agree with this stance. She later became an anti-nuclear activist and during the Vietnam War years supported conscientious objectors and agitated against the war, always by nonviolent means. There is disagreement within the movement about her stance on abortion, as on other issues such as loyalty to the authority of the church. It is clear that she opposed abortion, but she had had one herself and understood the pressures on women faced with an unwanted pregnancy. One lifelong Catholic Worker, who knew and worked with Day in his youth, writes that she did not campaign on abortion, never spoke, wrote or marched in favour of criminalising it; her opposition was 'quiet and personal', and without judgment of the women and their families who chose the procedure. Others argue that while she affirmed the rights and dignity of women, she also felt that nurturing life should be a priority for all adults, and sometimes required sacrifice.

The movement took its name from the *Catholic Worker* newspaper that Day founded in 1933, still publishing, but it is best known for its houses of hospitality, providing food and shelter to those in need. They are usually in poor urban neighbourhoods, although there are also Catholic Worker farms. Living simply on the land was particularly espoused by Maurin, who believed cities were an aberration.

The first live-in Catholic Worker house in Australia was started in West End, Brisbane, in 1982, by Ciaron O'Reilly and Angela Jones. Jim moved in soon after. In 1986, after it folded, Jim, Anne and others started a new Catholic Worker house in the same area. It was where they spent the early years of their marriage. In 2000, with a growing young family, they moved onto church-owned rural land in Dayboro, about an hour's drive north of Brisbane. A sympathetic priest had helped them obtain a peppercorn lease. It was a grazing paddock when they first arrived. They named it after Peter Maurin and in his spirit planted its shade trees, orchard and gardens, and kept their doors open, as much as they were able, to people in need while they raised their seven children.

Franz is their fourth-born. They named him after Franz Jägerstätter, a humble Austrian Catholic farmer executed by the Nazis when he refused to enlist in the German military. If quiet but unwavering resolve guided by conscience is what they hoped this son would learn from his namesake, Jim and Anne chose well. Franz left his parents' home as soon as he left school but has chosen to live like them, in voluntary poverty in a Catholic Worker house of hospitality, known as Dorothy Day House. In Greenslopes, on Brisbane's south side, the five-bedroom house belongs to a cousin but is managed by a real estate agent and let at market rent. Franz, as one of its six core residents, does a shift as a disability care worker to pay his share while also studying for a double degree in human services and creative industries. In between he helps keep the house open for people in need, growing food in the backyard to go towards feeding them, dumpster-diving for the rest. This is the activism of his everyday life, demonstrating a 'radical sense of community and a better way to live'.

Then there is 'reaching out to the world'. After a West Papuan refugee sought shelter with them, the whole household committed to keeping a weekly public vigil for West Papuan independence, a cause which till then didn't have much visibility in Brisbane. Another of their weekly commitments is running a 'Food Not Bombs' street kitchen. It's a demanding life, but Franz intends to stay in the Catholic Worker house, at least for the foreseeable future: 'It's a big part of my heart.'

His large, loving family are another big part, though Franz found it difficult, as he grew up in Dayboro, that they were so different from everyone else. One of the things that drew attention to them was Jim's old blue van, emblazoned on the back with the slogan 'Christians Against All Terrorism'. Another was his bare feet. 'We kinda stuck out in that small town.' All of the boys were bullied at school, Franz and his older brother Joseph getting the worst of it. Jim had no idea how bad the bullying was until his sons told him about it in recent years. He says he couldn't have lived differently, but he cries when he speaks of it, especially to think that he didn't do enough to protect them. 'Even in the local church people weren't very supportive of us,' Franz recalls. 'It was always a bit difficult, but Mum and Dad are both very stubborn. And we all got through, having each other.' And Franz too these days often goes about in bare feet.

Arrestable civil disobedience actions are new for him. At nineteen years old the Pine Gap trespass is his first, but by the time he gets to trial he'll have notched up another. He knows he will be seen as acting under the influence of his father, but he is quite clear about what motivates him to act: 'What is there not to hate about Pine Gap? Everything that it stands for and is just seems so morally wrong ... One of its primary uses is for drone strikes. These machines kill innocent people in foreign lands where we are not at war. I think this is a big driving factor in what angers me about Pine Gap. It is very hidden, not many people know or care about it, but a question that really, really needs to be asked … is: What is this base and what do we really gain from it?'

Before the Pine Gap action and until after the trial, Andy Paine and Tim Webb were also living in Dorothy Day House. Andy laughs when he explains he is a Catholic Worker but not a Catholic: 'There's an old joke about Catholic Workers – most aren't Catholic and most don't work!' He describes his faith as 'Christian anarchist'. He will seek out non-conformist churches as places of worship but he doesn't pledge allegiance to any denomination. When he was growing up in Mudgee, in country New South Wales, his parents were involved with the Salvation Army; he recognises that as an influence on his ethics. After leaving home he spent time with a Pentecostal Church, studying for a degree in theology, before coming into contact with the Catholic Worker Movement through Jim and his family. There he found the kind of community he was looking for – people who believed in an active God at work in the world, who were not looking for wealth, power or success, who held themselves accountable to their fellow human beings wherever they were.

When he jokes about not working, he's referring to wage labour. For close to a decade he has lived with less than $5000 a year in earnings (and no Centrelink), but his life is full of endeavour. He does reporting and reviewing for community radio stations in Brisbane and Melbourne, he writes a thoughtful blog, he plays music. Like Jim and Franz, he does a shift as a disability care worker to come up with rent and he also does his share of dumpster-diving – a way of looking our wasteful society 'square in the eyes and rejecting it as the only way of doing things'. It's also necessary in order to feed the large number of people who come through their doors in Greenslopes or who turn up at Food Not Bombs. In part due to being older – thirty when he is arrested at Pine Gap – his nonviolent direct action history is much more extensive than Franz's, and it includes a few arrests. His nomadic adventurism also takes him to where the need is. Andy thinks nothing of crisscrossing the country with just his guitar, beg-borrow-and-paying his way with his busking proceeds.

As he sees it, he has grown up with war. He had just turned fifteen when Australia entered the war in Afghanistan, so that accounts for more than half his life. He was still at high school in Mudgee as millions around the world marched against going to war in Iraq. His idealistic hopes soared briefly, then crashed when George W Bush's 'Coalition of the Willing' – including, of course, Australia – went in anyway. For Andy, becoming concerned about the role of Pine Gap in those wars was a logical next step.

Tim, like Franz, grew up inside the Catholic Worker Movement and there's a strong similarity between them. Not in features or colouring but both are introspective, shy, Tim perhaps more so; during the trial both wore their long curly hair tied back or else up in a knot; both are tall and their years of farm work have made them physically strong. Being raised in Catholic Worker households has bred in them another kind of strength and resolve, necessary for the kind of activism they undertake.

Until he was seven Tim lived in Hokianga in Northland, Aotearoa New Zealand, where his extended family owned a farm. When he was born, the fourth of eight children, his maternal grandparents and five of their nine children were living there, with their spouses and their children. The property is mostly steep hills and bushland, but the families cultivate together some ten acres of fertile river flats, growing most of the food they need. The ground is tilled using a horse and plough, weeded and maintained only with hand tools, no machinery or chemicals; cars are used only off the farm. Their houses are hand-built, each with a twelve-volt solar system, enough for a radio and lights; water comes from a fresh-water stream out of the hills; toilets are long drops or composting.

Apart from this commitment to simplicity, the families believe in 'loving your neighbour and the earth in a deeply Christian sense'. A neighbour is not simply the person next door, explains Tim, rather 'the person in need who is near you'. Hokianga and Northland in general are poorer than most other regions of New Zealand so they don't have to look far.

When his parents left Hokianga, moving closer to Auckland, it was mostly for better work and schooling opportunities but there were also lifestyle changes at the farm that his parents weren't sure they could sustain – like no screens of any kind. Both parents, but especially his mother, continued to instil in their children the value of causing as little harm as possible to themselves and other people. They made them understand that the luxury and convenience of Western lifestyles were paid for by the 'short, unbelievably miserable lives' of other people halfway around the world.

Despite these strong values, Tim says his parents 'never openly profess to be Catholic Worker – they humbly assume that they are not worthy of the name'. He disagrees: 'Catholic Worker is not some sort of exclusive elitist club, but the complete opposite. There is literally nothing you need to do to qualify except have a willingness to love your neighbour. So I think Mum and Dad could call themselves Catholic Worker.' Still, he can see where his parents have drawn certain lines. Catholic Worker activism is in part about having 'difficult people who are marginalised by society actually living in your home. Mum and Dad would often have these people visit, but not usually to live with them. They would go to protest marches and prayer vigils but they haven't been arrested, not that getting arrested is the goal but it does indicate a certain degree of radical thinking and living.'

Tim was just fourteen when, in 2008, a cousin of his, Sam Land, trespassed at the Waihopai base in New Zealand, that country's contribution to the Five Eyes global mass surveillance effort, as Pine Gap is Australia's. With two other Catholic Worker activists, Sam slashed an inflatable radome covering a parabolic dish antenna that the trio believed was contributing to the Iraq War. Their action made so much sense to Tim. For him, Jesus's teachings were clear: if people renounced violence, the world would be a better place. Tim read the press coverage of his cousin's trespass and trial with keen interest, learning about 'the shadowy side of the government' and its involvement through the base in conflicts around the world. He was disheartened by the attitudes of boys around him at school – they had an unrealistic, video game view of the world and of war, he thought – but the foundation of his own activism had been laid. A few years later he crossed the Tasman and met up with Jim and family and the rest of the Brisbane Catholic Workers. The Pine Gap arrest was his second, at age twenty-two, for an act of civil disobedience.

Alone of the group Margaret Pestorius is not a Catholic Worker, though she is a committed Catholic. 'I don't have voluntary poverty, I have voluntary extreme riches!' she jokes. Even when she is speaking about serious matters, her dancing brown eyes often urge you to laugh. She pokes fun at herself as an older woman, asking to be introduced as a 'direct action goddess'. She wears her straight hair, grey now, at shoulder-length, sometimes pulled back with a hair-tie, glasses, no make-up; dresses in comfortable shoes, mostly in shorts and t-shirts, the latter usually sporting a campaign slogan or the logo of one of her peacemaking organisations. More than in her dress though, her activist identity is expressed in her aura of energetic purpose.

Her 'extreme riches' mean, in part, that she comes from a well-off family and has inherited wealth; she also practises her profession – social work, therapy – and gets paid for it. She questions, gently, her friends' dedication to voluntary poverty. She sees it as so time-consuming and its framework of action – the houses of hospitality, involving a tremendous effort in feeding and sheltering people – as somewhat narrow. 'When you're doing community-based organising, your relationships have to be broad.' On the other hand, they are the people in the peace movement who undertake direct actions, time and again.

In many ways Margaret seemed destined for another life. At All Hallows' in Brisbane the Sisters of Mercy taught her piano and violin, and she sang in their choirs. This was formative for her. Later she was sent to the elite Brisbane Girls Grammar. Music became her primary interest and she went on to study at the Queensland Conservatorium and the Royal Conservatorium of The Hague in the Netherlands. Returning to Australia she began playing professionally. Through one of her fellow classical musicians, she came into contact with the

Melbourne Rainforest Action Group. They introduced her to the demands and rewards of nonviolent direct action, and for many years she turned away from music. With them she took part in thirteen actions on the Yarra River, when activists went into the water by whatever means they could, including swimming, to blockade ships importing tropical rainforest timber into the city. More recently, along with many other environmentalists, she has turned her attention to coal. She has been active in Stop Adani Cairns and Friends of the Earth FNQ, and a member of the Greens, but central in her life, with an almost daily quantum of tasks, is her work for peace.

Apart from direct actions, she is building 'organisational infrastructure' for the movement. She has created a company called Australian Nonviolence Projects whose major activity is to support the campaigning instrument Wage Peace, employing a worker (a rare thing in the peace movement) for ten hours a week. The company has five objectives: to join people and projects, to amplify messages, to maintain digital assets (their online and social media presence), to reveal pathways for action, to build capacity at every level. As if this list is too business-like, Margaret adds another: 'And sometimes to create a spark.' She sees their target as 'the average Jo' – she specifies the female name – aged forty to eighty. She feels frustrated, though, by the resistance of many older people to learning the skills necessary for contemporary campaigning – the digital, organisational and management skills she has developed in her professional roles. Face-to-face activism suits these people better but it tends to be sporadic and is not the only thing that's required.

A separate arm of her company is Beyond War, devoted to fundraising: 'The idea is to use my money to get other money' that she directs to projects like West Papuan independence actions. She oversees and nurtures all this by phone and laptop from the Peace by Peace House, the large old Queenslander where she and her husband, Bryan Law, lived together, often with other people, and raised their son, Joseph.

Margaret became friends with Jim Dowling during the court case that followed the 2005 trespass at Pine Gap, in which Bryan was also involved. He was preparing to stand trial for another pro-peace, anti-war action, when he died in 2013. He was fifty-eight years old. Margaret speaks of him often, including his death, in a seemingly matter-of-fact way. How much she esteems him, and misses him, can be seen in part through her efforts to maintain his *Peace by Peace* blog and upload other writings by him and about him to an archive site. Perhaps it can also be seen in her attention to the broader human experience of grief. It laid a foundation early in her own life. She was just three when her father died from cancer, but it took her a long time to recognise this loss and finally grieve for it. Then after Bryan died, apart from her own grief, she had to watch their son go through his. She saw how it added to the legacy of loss running through her family, with her mother having been orphaned at age fifteen. Margaret has taken this experience into her work as a therapist and into her political understanding and actions. In Cairns she has been involved in Frontier Wars ceremonies that are about 'drawing people into lament, speaking the truth of something terrible in our national story'. Lament is also a big part of how she thinks about the trespass action at Pine Gap – lament for the lives lost at the end of its kill chain.

~

How do you talk about Pine Gap, the Peace Pilgrims have asked themselves, as I have – about what people do there, in whose service, and what that means in the world – in a way that isn't just dismissed as uninformed, cranky, fanciful or hopelessly naive? And in a way that makes it real. Fifty years of secrecy and highly complex technology make it seem so unreal we can scarcely think about it.

I began to read, not only the work on the base by academic researchers and investigative journalists, but in the military sector. An article about military logistics by an Australian commanding officer led me to a 2018 paper by the US Joint Chiefs of Staff that advances a new intellectual framework for what they do. The Chiefs argue that the 'peace/war binary' is obsolete. Instead they want to see military and civil activities aligned along a 'continuum of cooperation, competition below armed conflict, and armed conflict'.

The campaign is worldwide, continuous into the foreseeable future and framed entirely to project US power. The Chiefs seek to maintain the existing international order that has been challenged by 'emerging and resurgent global powers, aspiring regional hegemons, and non-state actors' competing with the US. Nowhere, though, in the document's forty or so pages is there a vision for, or even any reference to, striving for 'friendly relations among nations based on respect for the principle of equal rights and self-determination of peoples', as put in the United Nations Charter to which both our nations are signatories.

The Pilgrims live so far off the Chiefs' competition continuum it's not funny, but I expect many Australians would be as perturbed as they are by the unvarnished insight the Chiefs provide into the thinking of our powerful ally. This is not the future: this 'forever war' is happening now, and Australia, through its government and military, is embracing it despite little awareness of this in the general community. When I talk to people around me about what I've learned, they are truly dismayed. There are nano-explosives, with ten times the force of conventional explosives, that can be mounted as warheads on a wide variety of small drones. 3D printing will allow the large-scale manufacture of cheap, small smart drones – tens of thousands in a day with only one hundred printers is already possible. The US and Chinese militaries are already working on launching large numbers in minutes. Think of them as expendable rounds of ammunition – coming at you in swarms. They won't be controlled, even remotely, by pilots. Designers are working on autonomous navigation and targeting, drones taking over once the commander has decided where they will be sent and who or what will be targeted. Targeting is the technically more challenging problem, but applications being developed commercially for driverless cars and AI-driven cameras will be useful for improving the drones' 'hunting capability'.

In March 2018 the Royal Australian Air Force convened its biennial Air Power Conference, titled Air Power in a Disruptive World. Its sponsors Rolls Royce, L3 Technologies and Boeing make their money from the latest generation weapons systems. They would have every reason to be pleased with Australia's ambition to enter the top-ten weapons export league within the decade. In the meantime we are also good customers. Then defence minister Marise Payne's speech to the conference was long on her government's 'heavy' investment in the technology of war, some $195 billion over ten years. It was entirely devoid of reflection about the resort to lethal force, the grave weight it is due, the ethical challenges presented by what is described as disruptive technology but accepted, it seems, as prescriptive – entirely inevitable.

More than one conference speaker talked of the core need to accelerate the decision cycle – to be able to make best use of the vast amount of data which Pine Gap and its network suck up. Think of artificial intelligence as our 'intellectual partner' in this, urged one. Digital natives, growing up with Xbox games and the like, 'need to run the show'. Trust needs to be built, not in one another, but in AI. Russia has declared that AI is the future and that 'whoever becomes the leader in this sphere will become the ruler of the world'; China is unambiguous about its AI plans and already has the advantage of military–civil fusion; when 'our adversaries' are going down this path our ethics could be the 'handcuffs' holding us back. Another speaker, referring to the 'grey area' of lethal force short of outright conflict, asked whether hybrid warfare could be used for good. But what good, whose good? He didn't go there.

The picture of this future is dark – for all the shiny imagery of its PowerPoint presentations, with not a dead body in sight, nor devastated city, nor vast refugee camp. We will live in an intensively militarised world, with an ever-ready prospect of rapid resort to lethal force on many fronts. Pine Gap and the network it is part of are integral to this future.

At the time of the drone strike on Shadal Bazar, Obama was right at the end of his presidency. He had entered politics with a strong focus on nuclear disarmament and was awarded the Nobel Peace Prize for his efforts. Yet, while he made some gains in non-proliferation, they came at the cost of modernising and thus entrenching the US nuclear arsenal for decades into the future. He also established an unenviable record for the secret killing of individuals to advance US security goals.

The drone attacks were justified under US law by a joint resolution of both houses of Congress, passed in the wake of 9/11. The *Authorisation for the Use of Military Force* (AUMF) allowed the president – at that time George W Bush – to use 'all necessary and appropriate force' against the perpetrators of those 'acts of treacherous violence' and the nations, organisations or persons that he determined had helped or harboured them. This was not limited to retribution for the attacks on New York, itself arguably illegal under international law because lethal force is permissible 'strictly to prevent the imminent loss of life'. The authorisation also extended to preventing 'any future acts of international terrorism against the United States by such nations, organizations or persons'. The legal test for anticipatory self-defence, outside the context of active hostilities, is that the threat must be 'instant, overwhelming, and leaving no choice of means, and no moment of deliberation'. There are very few situations in which that test would be met. Meanwhile, in America's 'forever war' the battlefield, or rather hunting ground, has become the world, an assault on the principle of territorial integrity that underpins state sovereignty.

'There is no time limit, no sunset clause,' wrote former Australian prime minister Malcolm Fraser in 2014 about the AUMF, 'there is no geographic limit to the presidential power ... For the first time the power was directed not at nations but specific organisations or persons. The President himself is the sole arbiter. The power is unending because it involves the prevention of future attacks.' Fraser's voice, in a book tellingly titled *Dangerous Allies*, had some cut-through on this issue, in Australia at least. He had seemed, as Minister of the Army and then Minister of Defence, to unquestioningly oversee Australia's military involvement in the war in Vietnam, but by the end of his life – after Australia had followed the US into Afghanistan, into Iraq – he was blunt. He challenged Australians to recognise the paradox contained in our strategic dependence: 'We need the United States for defence but we only need defence because of the United States.' This wasn't embraced by everyone, far from it – most Australians still cleave to the American alliance – but it did prompt debate.

Under Obama the AUMF was used to cover going after organisations that did not even exist in September 2001, including Daesh. Air strikes, largely by drones, targeting Pakistan, Somalia and Yemen, numbered 563 during Obama's two terms, compared with 57 under Bush. Civilian casualties in these countries, which were outside active war zones, were estimated in the several hundreds.

Donald Trump took over in 2017, handing off to the Pentagon a lot of the oversight that Obama had insisted on and walking back his limited transparency measures. In Afghanistan the number of strikes approached levels last seen during the 2009–12 surge, with close to two thousand strikes, by plane and drone, in 2018; in 2019 the figure soared to almost seven thousand. There was a marked decrease in Pakistan in 2018, down to 5 strikes; 45 in Somalia (63 in 2019); 36 in Yemen (9 in 2019).

The casualty numbers are a drop in the ocean of civilian deaths under American bombs since the Second World War, in Korea, Vietnam, Laos, Cambodia, Afghanistan, the Sudan, the Balkans, Iraq and now Syria. Indeed, the history of bombing, and not only American bombing, has been a history of mass civilian casualties – the 'customary law' of air wars. Targeted killings by drone, with their comparatively limited toll, may seem to be a more 'humane' alternative, even if mass killing by nuclear strike remains firmly on the table.

Under international law for the conduct of war, the tests for the use of weaponised drones are the same as for any other weapon. Parties are obliged at all times to distinguish between civilians and combatants. A wilful attack on civilians with no military objective is a war crime, terrorising a civilian population as a primary objective is a war crime, an indiscriminate attack is a war crime, no matter what the weapon used. But air power, from whose cold logic drones are an outgrowth, has presented particular challenges for adjudication, among them one that goes to its very nature – the 'diffusion of responsibility between those choosing the target and those pushing the button'.

That the aeroplane would change the nature of war was foreseen with extraordinary prescience by the great Indian writer and leader Rabindranath Tagore, whose life and thought, particularly in this domain of militarism, has been drawn to recent Australian attention by Barry Hill's epic *Peacemongers*. Tagore first flew in an aeroplane in 1932, to Persia. At that distance above the earth he experienced the loosening of its hold on heart and mind. Looking down, he asked, 'Who is kin, who is stranger?' How would one know? Extrapolated to now, it is among the questions that haunt targeting decisions for the deployment of drones and the debate around their legitimacy. Who is a civilian, who a combatant or direct participant in hostilities? How do we know? Where there is doubt, civilian status must be presumed.

The military and strategists though, on the whole, have embraced the sheer efficacy of bombing's death-dealing. In the beginning, in the early twentieth century, its victims were the 'barbarians': air power's utility in the colonies, allowing 'control without occupation', kept it out of the international agreements on the laws of war. Then in the Second World War Europeans turned bombs against themselves – whole cities laid to waste before the apotheosis was reached by the US in Japan. The Allies' own record removed aerial bombing from consideration in the war crimes trials at Nuremberg and Tokyo, and they worked to limit the protection of civilians from bombing under the Geneva Conventions. Onslaught from the air, in which 'the mass murder of civilians has been central', has since become the American 'way of war' even though, in conflict after conflict, victory has remained elusive.

The particular legal concern with drones has been the way they lower the threshold for the resort to lethal force. Because they make it easier to kill without risk, there is a temptation to push the legal limits on who can be killed, and under what circumstances. Outside the context of armed conflict, experts say that drone killing is 'almost never likely to be legal' and should be investigated and prosecuted as an extrajudicial assassination. Transparency and accountability by states for targeted killings are essential to their compliance with international law obligations. To date this has been very limited in the US. In Australia, in relation to Pine Gap's activities, it has been non-existent.

For the general public, civilian deaths and injuries as a result of targeted killings, despite their relatively small number, have been the primary focus of controversy. That these dead are people whom any legal process in liberal democracies would presume to be innocent only heightens the misgivings about what Obama described as this 'different kind of war' waged by the US post 9/11. In drone 'warfare', no longer do combatants face one another; rather it is man-hunting, high-tech predators going after low-tech prey, a policing operation without any of the other tools of policing, only 'shoot to kill'. It is without risk to the remote killers (completing the promise of the bomber plane) but in 'depriving the enemy of an enemy', the warrior ethos and the ideal of heroic sacrifice are profoundly troubled. The kill is transformed into an execution, the soldier into a 'mere executioner', which has its own psychic costs. This asymmetry prompts response with another asymmetry – the suicide bomber. And so the deadly cycle goes.

Unintended deaths in drone operations also test the claims of 'surgical precision' that supposedly make these attacks more humane. They undermine too whatever gains the attacks hoped to achieve in counterinsurgency terms, with every dead non-combatant representing 'an alienated family, a new desire for revenge, and more recruits for a militant movement that has grown exponentially even as drone strikes have increased'. This comment was made in 2009 about drone killings in Pakistan. It is remarkably similar to a British officer's assessment in 1923 of the perverse effects of aerial bombardment in Iraq: '[T]hese attacks bring about the exact political results which it is so important, in our own interests, to avoid, viz. the permanent embitterment and alienation of the frontier tribes.'

~

At one end of the chain, the drone strike on Shadal Bazar; at the other, America post-9/11. In the middle: the base at Pine Gap. Nobody doubts the capacity of the base and its network to pry into every corner of the globe, including into Australian lounge rooms if it wants to. Hollywood has softened us up for that. The critical consideration – for the Peace Pilgrims, for me, for Australia and its place in the world – is the military application of that snooping, tying it, in particular, to covert operations in countries that Australia and America are not at war with.

The series of reports about Pine Gap by Richard Tanter, Des Ball and others, published by the Nautilus Institute, provide meticulously detailed analysis of the several generations of technology installed at Pine Gap, giving a clear picture of the evolving capabilities of the base and their significant expansion.

When the Christians Against All Terrorism were preparing for their trial after trespassing at the base in 2005, they turned to Richard and Des for help. They wanted them to provide expert evidence of the base's involvement in the Iraq War. Richard and Des had known about Nurrungar's early warning of Scud missile launches during the Gulf War and they knew Pine Gap had taken over this role, but they had no hard information about a broader signals intelligence (SIGINT) contribution to the war. SIGINT is Pine Gap's main activity. Richard and Des started to look into it more closely and became convinced the Christians were absolutely right.

They later consolidated their understanding of Pine Gap's military contribution by matching their technological analysis with a careful tracing of Pine Gap personnel. It involved some great sleuthing, trawling the internet for any clue on who these people are, how many they are, posted to what service, what they did before arriving at Pine Gap and the role they went to at their next post. These traces allowed conclusions to be drawn about what their activity at Pine Gap would most likely have involved, and in some cases no surmise was necessary. The footnoted references are to sources such as the telephone directory of Buckley Air Force Base in Colorado and to surprisingly unguarded professional profiles on LinkedIn.

In the profiles former staff referred, for example, to providing 'collection support to over 25 combatant units'; to analysing '80,000 real-time signals of interest' and forwarding reports 'regarding combat, strategic, and tactical intelligence'; to analysing '53,000 threat reports', directing '700 hours of analysis to 250 Special Operations Forces missions', and providing data to 'military operations in Afghanistan'. This last was from an analyst also involved in supporting 'Personnel Recovery' – evacuating US wounded, rescuing downed pilots. Another analyst had worked on 'providing situational awareness to United States and Allied policy makers and armor forces worldwide'.

Added to this evidence from Richard and his colleagues is the memoir of David Rosenberg, an American electronic intelligence analyst who served for eighteen years at Pine Gap. He was not a soldier but the context of his work was distinctly military. He writes of his service being directly involved, after 9/11, in the hunt for Osama bin Laden, intercepting his communications. As President Bush geared up for his global war on terror, Rosenberg says analysts in many agencies were 'assessing Afghanistan's weapons systems and communications networks' while 'those of us in the eavesdropping intelligence community prepared for the subsequent bombing campaign'. Elsewhere, Rosenberg has been untroubled by the association of Pine Gap with drone killings, for example, telling the ABC in a 2014 interview: 'If the US military could make use of whatever assets it has in place to help the drone program, then the US military will of course do that. Whatever signals Pine Gap can collect that could be of use to the governments would certainly be passed on.'

Former US drone operator turned whistleblower Cian Westmoreland has described some of what happens at the other end, for instance, the way the work is spread:

> You have different countries doing different things all working together. You have stations in Great Britain and the Australians would be working with the Americans and the British. It's collaborative, and it's really hard to say 'the Australians are responsible for this' or 'the British are responsible for that'. Everybody is working together and if the Australians were involved in one piece that happened to be used in a strike, they're essentially complicit with whatever the end result is.

Des Ball, however, learned from government sources that at least one attack in Yemen, a strike against an Al Qaeda target on 3 November 2002, relied partly on signals intercepted at Pine Gap. While this was the only specific incident he could point to, he fully expected that since 'the war on terror' had been declared, 'hunting down terrorists by monitoring their mobile phone and sat phones has been one of the highest priorities of these operations'.

The involvement of Pine Gap in this 'new phase of warfare' changed Ball's mind about the base. Australian sovereignty had always been a prime issue for him, and he had always been critical of the unwarranted secrecy surrounding Pine Gap, as well as its essentially American character and purpose, its role in nuclear war-fighting plans, and its status as a likely nuclear target. In earlier decades, though, he had supported it, on balance, for its necessary role in monitoring nuclear and other advanced weapon systems in our region – intelligence that was critical for monitoring various arms control agreements, not collectable, he said, in any other way. He also supported the use of its satellites 'for picking up things closer to home – in other words, political and other developments in our neighbourhood'. The drone program had profoundly challenged that position: 'I've reached the point now where I can no longer stand up and provide the verbal, conceptual justification for the facility that I was able to do in the past. We're now linked in to this global network where intelligence and operations have become essentially fused and Pine Gap is a key node in that whole network, that war machine, if you want to use that term, which is doing things which are very, very difficult, I think, as an Australian, to justify.'

The questionable legality of drone assassinations was on the minds of at least two Australian parliamentarians after reports surfaced in 2014 of the deaths of two Australian citizens in Yemen. They had been killed by a Hellfire missile fired from an unmanned US drone on 19 November 2013. 'There was a target that the Americans were after and these people were with that target,' said then foreign minister Julie Bishop.

In June 2014, in the House of Representatives, Labor MP Melissa Parke expressed her disquiet about

> a death sentence ... pronounced and carried out from a drone controlled by people we cannot know who are not accountable to any legal process we would recognise and whose decisions we are being asked to accept on trust after the fact. And this occurred in a country with which we are not at war and which is relatively far from

> any area where our military forces are directly engaged. Can such strikes be reasonably considered acts of war or are they more properly to be regarded as instances of extrajudicial killings or war crimes?

Parke was in a better position than many to understand the issues: before entering the parliament she had worked as a senior international lawyer for the UN in Kosovo, Gaza, Lebanon, Cyprus and New York.

In the Senate in May, the Greens' Scott Ludlam had tried to elicit an Australian Government view on the legality of such a strike, under international law. Attorney-General George Brandis was 'not aware that the Australian government has a view of legality of these matters'.

Ludlam attempted to raise the role of Australian officials at Pine Gap doing these 'security assessments of individuals who are then assassinated'. Was the attorney-general confident they were 'immune from future accusations or allegations of assisting war crimes?'

'I am very confident, without admitting any of the assertions or premises of your question, that ASIO and its officers operate in accordance with Australian law,' said Brandis.

'What about international law?' asked Ludlam.

'And in accordance with international law.'

How did he know, had he sought legal advice? Ludlam persisted.

'I receive advice, including legal advice,' said Brandis, refusing to go beyond that statement.

The Australian Government, whichever party is in power, is on the whole a perfectly behaved alliance partner. The blanket refusal of information and comment is just as the US wants it, as was made explicit in the documents from the massive archive leaked by Edward Snowden.

'Support to Military Operations' is a clearly stated component of Pine Gap's mission in its top-secret *Site Profile*. The base's cover term is 'Rainfall', which I find especially galling for this military operation in the desert, where falls of rain are rare enough to be communally celebrated as the life-giving events they are. Activities at Rainfall include 'the collection, analysis, and reporting against assigned targets' of intelligence, says the profile. These days, communications intel – phone calls, emails – accounts for eighty-five per cent of what is collected, according to another classified document.

All of this, however, is supposed to masquerade behind a bland facade. The *Site Profile* provides an unclassified mission statement for the base – what the Chief of Operations should say about it, 'without further expansion or explanation'. It is the usual brief, almost meaningless fare: the base is 'a joint U.S./Australian defense facility whose function is to support the national security of both the U.S. and Australia'. The profile goes on to advise: 'Avoid any implication that this statement is only a sanitized portion of a larger classified effort.'

Another top-secret document gives a general definition of Australia's role in that effort: 'manning of the operations floor' at Pine Gap, 'a site which plays a significant role in supporting both intelligence activities and military operations'. A lot of this document is concerned with China as an intelligence target: 'Increased emphasis on China will not only help ensure the security of Australia, but also synergize with the U.S. in its renewed emphasis on Asia and the Pacific.' It ends with a big tick for Australia: 'Problems/Challenges with the Partner – None.'

Important wording that the alliance partners rely on in relation to Pine Gap's activities is 'Full Knowledge and Concurrence'. The *Site Profile* makes a point of its usefulness: 'FK&C is important so that the Australian Government can assure the Australian Parliament and the Australian people that all activities conducted at JDFPG are managed with the GOA's full agreement and understanding.'

What looked like an official Australian scruple about this came from Stephen Smith as Minister for Defence in the Gillard and Rudd Labor governments. In June 2013 he updated 'the parliament and the Australian people', for the first time in six years, on the 'joint facilities'. His ministerial statement followed increased cooperation and engagement with the US, including the US Marines 'rotation' in Darwin, with 1150 marines expected in 2014, building to around 2500 in the coming years.

In relation to Pine Gap, Smith covered the familiar ground of its history. Under the heading of 'evolving role', it was still all about the old story – monitoring arms control and disarmament agreements and providing early warning of ballistic missile launches. His scruple crept in when it came to an explanation of what 'full knowledge and concurrence' means:

> 'Full knowledge' equates to Australia having a full and detailed understanding of any capability or activity with a presence on Australian territory or making use of Australian assets. 'Concurrence' means Australia approves the presence of a capability or function in Australia in support of its mutually agreed goals. Concurrence does not mean that Australia approves every activity or tasking undertaken.

Malcolm Fraser was specifically critical of this as an imagined way out for Australia from the consequences of American actions:

> The words 'full knowledge' and 'concurrence' means [sic] that we approve the use of Pine Gap for targeting American drone killings. How can we argue otherwise? What nation would allow a base on its territory to conduct missions about which it has full knowledge, but of which it does not approve? It is simply an impossible scenario, a contradiction in policy.

Of course Smith, like ministers before and since, made no mention of Pine Gap's involvement in drone killings. By never talking about the issue, they never have to confront the questions of consequences and legality. Fraser described the American legal justification – its reliance on the *Authorisation for the Use of Military Force* – as 'tenuous at best'. He couldn't see that the authorisation would stand up in the International Criminal Court, leaving US personnel 'vulnerable' to prosecution. Not that the US would allow that, as Trump and his team have made perfectly clear. Australia, however, is a state party to the Rome Statute that established the court, and Australian personnel involved in drone killings, in Fraser's view, have even less to fall back on than the Americans: 'Australians operating at Pine Gap are doing so without any authority and could be equally liable.'

Human rights advocacy groups and international law experts, as well as the UN special rapporteurs on the use of drones in counterterrorism operations, have all repeatedly emphasised the obligation on states to be transparent and accountable in this domain. A first step, they say, would be public access to the relevant information, such as the criteria for targeting and the authority that approves killings. The Australian Government has been entirely impervious to this advice.

In early 2019 Defence Minister Christopher Pyne, a couple of weeks before announcing that he was quitting federal politics, made a new ministerial statement about the bases. He emphasised that Australians hold key decision-making positions at Pine Gap and have direct involvement in operations and tasking. He also said that the government has full oversight of activities undertaken there and that they are in accordance with Australian and international law. This was asserted without any development of what the issues might be, saying only that 'much has been theorised' about the base's involvement in US operations against terrorism. Even the head of the Australian Strategic Policy Institute, who provided a generally glowing assessment of Pyne's statement, described this as 'rather coy'; surely the minister meant 'US and Australian' operations against terrorism. 'Is the implication that we gather intelligence on terrorism but do nothing with it?' he asked.

Pyne also explained the 'full knowledge and concurrence' policy, reflecting the same nuance as Smith on approval, before going a little further on 'concurrence':

> It doesn't mean that Australia approves each and every activity or tasking undertaken; rather, it means that Australia agrees to the purpose of activities conducted in Australia and also understands the outcomes of those activities. But I can assure the Parliament and the Australian public, we maintain appropriate levels of oversight for the activities undertaken. Importantly, concurrence also means

> that Australia can withdraw agreement if the government considers that necessary.

Given this apparent assertion of sovereignty, while hardly expecting answers, I did put some questions to the minister, as well as to the shadow minister, Richard Marles, whose statement in reply in the parliament 'happily' supported 'every word' the minister had said.

I asked the minister to explain how the provision of targeting data by Pine Gap for lethal drone strikes by the US in countries that Australia is not at war with conforms to Australia's obligations under international law. Specifically, I asked, what are the international law principles and rules that govern this kind of operation? What is the Australian law that authorises Australian personnel at Pine Gap to be involved in this kind of operation? Has the government sought specific legal advice on the liability of Australians involved? If so, what is that legal advice?

Pyne had said in the statement that 'the intelligence we produce together has saved lives'. Given that it has undoubtedly also taken lives, including those of civilians, I asked what the minister had to say about the accountability and responsibility of Australia on that score.

A Department of Defence spokesperson eventually replied: 'Consistent with longstanding practice, the Government does not comment on intelligence matters.' There was no reply whatsoever from Marles.

~

I can feel quite intimidated by the depth of what I don't know in all of this. It would be easy to retreat once more into the peacefulness I experience at home – where I can look up with assurance into the endless blue of desert days, where the only buzzing comes from flies, where at night a mopoke might call into the stillness or lift with a flurry of wings as I step outside. But it feels dangerous to stay quiet, to let technological development and its deployment run so far ahead of general understanding of what it is capable of and actually doing, let alone of its implications for the kind of nation and place in the world that we are said to be defending.

When the Peace Pilgrims tried to bring evidence into court of the lethality of what Pine Gap does, the Commonwealth's legal counsel repeatedly objected. The testimony was offered by Richard Tanter; the objection was that the knowledge was based on his own research: 'What can the Crown do with it, Your Honour? We can't cross-examine in the dark.' I had to laugh. Exactly the Pilgrims' point.

TRESPASS

Jim made his first trip to Pine Gap in 1987, long before the Pilgrims came together or were even thought of. He bought an old council bus to bring twenty to thirty people with him from Brisbane. Joining people from all over Australia, they formed into affinity groups to do actions, ten to twenty people in each. Jim's affinity was with Catholic Workers like himself, about six of them, all from the West End Catholic Worker house of hospitality.

The initiative for the convergence had come from the Alice Springs Peace Group, which had been active for a decade. They had 135 local members but, in face of the apathy if not hostility of the rest of the town, they relied on the Australian Anti-Bases Campaign Coalition to build support. Around five hundred people took part – one of biggest gatherings Jim had been involved in. Of all the protests at Pine Gap, only the Women's Peace Camp in 1983 was bigger, attracting around seven hundred women, 'the biggest, most lifestyle-diverse, grass-roots gathering of women in the nation's history'.

The 1987 campaign was planned under the banner 'Pine Gap: Closed by the People'. The Peace Group hoped to literally close the base, even if only for a short while, on 19 October, the end date of the second ten-year agreement on its arrangements between the US and Australia. They were concerned that the base undermined Australian sovereignty; they wanted to see the country develop a non-aligned foreign policy, strengthening its relationships with other non-aligned nations, and taking a firm stand against the nuclear arms race and militarism generally. They also feared the possibility of a nuclear strike on the base, with Alice Springs just nineteen kilometres away. While the Australian Government refused to discuss this openly, Des Ball had long been convinced that Pine Gap as well as the base at North West Cape and possibly the one at Nurrungar (not closed till 1999) would be priority targets in a Soviet nuclear attack. Paul Dibb, a senior policymaker and intelligence officer and later emeritus professor of strategic studies at the ANU, agreed, writing in 2018 that it was 'remiss, to say the least, of successive Australian governments not to provide even the most basic civil defence measures against nuclear attack on those facilities – despite the fact that they were advised to do so'.

Jim and his Catholic Worker friends did a couple of actions at the front gates. Later they sneaked around the back, went through the cattle fence that makes the outer boundary, and got to the inner boundary – three-metre-high chainlink fences, two of them, ten metres apart, right up close to the radomes and the operating area. They didn't go over the fences, or through them, didn't intend to, but they did get arrested. Jim starts to laugh, remembering it: 'We actually dressed up as cockroaches, being one of the few creatures to survive in nuclear war, they reckon. Pretty stupid, but that's what we did.'

I can see him, short, wiry, intense, curly hair and bushy beard, prancing at the fence, feeling a bit foolish but at the same time defiant, and I bet he was laughing. In court he was often rather gruff but among family and friends he is warm and good-humoured. All of the Pilgrims are, frequently laughing together as they gathered at the bar table to conduct their own defence. Part of it might have been a nose-thumbing to the formality of the court and the heaviness of the Commonwealth presence; part of it might also have been nerves; but part of it was their practised, spirited playfulness, their dance to power.

The arrest was far from Jim's first. He knew what to do. He never pleaded guilty, not for a civil disobedience offence, but admitted all the 'particulars' so he could get the hearing over and done with quickly: 'Things were a bit simpler in those days, magistrates gave you a lot more leeway.' He was fined, perhaps $100. He didn't pay: 'I probably would have gone to gaol for that, a few days or weeks, whatever. In the '80s and '90s I went to gaol many times for things like that. Twice I got a three-month gaol sentence straight off.' Things changed in Queensland after 2000, when they brought in SPER, the State Penalties Enforcement Register, in an effort to stop gaoling people for fine defaulting. They've mostly succeeded: they do things like suspend Jim's driver's licence from time to time, but it's not enough to wipe from his SPER list about $11,000 in fines.

His next arrest at Pine Gap came during the convergence of October 2002. The Anti-Bases Campaign Coalition were again one of the main organisers alongside local activists (the Alice Springs Peace Group had dissolved); also involved were Ozpeace, the Australian Peace Committee, the Anti-Nuclear Alliance of WA, Friends of the Earth Australia, the Medical Association for Prevention of War, and the National Union of Students. I'm struck by the dense network of civil society organisations that exists in this domain, even as they come and go or morph into others that are more enduring. The Pilgrims are small, fluid, and relatively recent in the context.

The original focus for the campaign in 2002 was the 'Star Wars' missile defence system, which had been tested over the Pacific in July of the preceding year. Denis Doherty, national coordinator of the coalition, had described the test as 'the most devastating blow to world peace seen in recent years'. Then came 9/11 and George W Bush's bewildering response to it – the impending US invasion of Iraq which Australia would actively support. Meanwhile, tacit support was being delivered through Pine Gap, as the campaign press release put it: 'At the moment, satellites monitored by Pine Gap are preparing the target logs for the US military to use in an attack on Iraq. Therefore, even as the Australian parliament debates our participation in a potential war with Iraq, Australia is already involved through the use of the Pine Gap facilities.'

Jim had seen the anti-war movement diminish by then from its peak in the '80s, when real fears of the Cold War turning hot had seen People for Nuclear Disarmament leading large protests in the streets, and Western Australia had put Jo Vallentine into the Senate on a nuclear disarmament platform. The huge international protests against the war in Iraq, when millions around the world took to the streets – and were ignored – did not take place till February 2003. In the preceding year, though, among a dedicated core there was no waning of commitment. Doherty and fellow organiser Gareth Smith arrived in Alice two months ahead to prepare for the October rally, so they were here on the first anniversary of 9/11. In the *Alice Springs News* Erwin began his interview with them, published on that date, with a respectful acknowledgment:

> Alice Springs people have more cause for reflection today than most Australians about the massive terrorist attacks in the USA one year ago. Our big American population will be feeling deep sadness.
>
> But there are grave questions about why Pine Gap, which brought them here, and the global intelligence system of which it is a vital part, were unable to prevent the carnage of 'Nine Eleven'.
>
> And many people are asking to what degree Alice Springs is exposed to terrorist attack because of the presence of the base.
>
> The town council looks at it mainly from an economic angle: Pine Gap's 'contribution has been estimated as $37–$52 million per annum,' the council says, surprisingly unable to come up with a more precise estimate.
>
> On the eve of US President Bush's speech to the UN about the case for war on Iraq, with unequivocal support from the Australian government, peace activists are arriving in The Centre for a large anti Pine Gap protest next month.
>
> Judging from similar events in the past, some protesters will be from the extremist fringe, but the core of the movement includes mainstream people putting their money, their freedom and – at times – their lives on the line for an ideal.

Erwin was at pains to overcome widespread antipathy in town towards visiting protesters, at least to the point of getting his readers to listen to what they had to say. He reported that Doherty, fifty-seven, was a part-time teacher in Sydney, devoting the rest of his time to the Anti-Bases Campaign; Gareth Smith, sixty, liked a laugh and was a full-time school psychologist in Byron Bay. A former UN volunteer in Timor-Leste, with a certificate of thanks for his service signed by John Howard and Alexander Downer, he also had against his name a string of civil disobedience convictions, all committed in the cause of peace. He explained why: when Australia was sitting on its hands as Indonesia was slaughtering East Timorese, he had organised a campaign of letters and petitions. The media all but ignored them. But when he and three of his mates risked their lives climbing to the top of Parliament House and spray painting 'Shame, Australia, Shame' on its roof, that made national news.

So, civil disobedience would most decidedly be part of the Pine Gap rally.

Smith and Doherty were expecting about five hundred protesters. In the end around three hundred came, mostly from Sydney and Melbourne, and others from Brisbane, Adelaide and Perth. The organisers suggested about forty per cent of them were people with 'kids, a job, a house and a car', another forty per cent students, the remaining twenty per cent people on the dole. Jim belonged to the first category – he came with his whole family, Anne and the six children they had then, travelling in the old blue van, leaving an aroma of fish and chips in its wake. Looking at him, with his bare feet and worn second-hand clothes, people might have thought 'feral', 'dole bludger', but they would have been wrong. His lifestyle is governed by a personal morality that is traditional in many ways: 'Drugs, booze, promiscuous sex, all that stuff, it's just irresponsible really.' He holds strongly to his responsibilities, in his marriage and to his children, to other family and friends, to people in need. He works for what cash is necessary – for 'where we plug into society, school fees, telephone bills, luxuries, what I see as luxuries anyway' – and otherwise labours on their small farm for his family's subsistence.

The police were very much trying not to arrest people then, as Jim tells it. He was doing his best to provoke them. This was not larrikinism, not acting out against authority for the sake of it: 'It's going to put a spoke in the machine, as Dietrich Bonhoeffer called it,' says Jim, referring to the German theologian and Nazi resister. 'The idea is to do a more serious action, actually disturb the war machine, rather than just make phone calls, or send letters, or hold placards in the street. If thousands did it, we could end the war, pretty quickly' – he hesitates, realising the enormity of the claim – 'possibly.'

He joined with Denis Doherty, whom he'd known for thirty years, blocking the base gates or trying to get through them. 'They kept dragging us away.' Arrests, and most of the media coverage, eventually came from a melee at the gates. A 23-year-old woman was charged with possessing an offensive weapon – paint-filled balloons. A flamboyant activist known as Captain Starlight – who favoured miniskirts, sequins, hot pink – was also arrested, after a police officer threw him to the ground and pushed his face into the dirt. 'I asked what I was being charged for and they couldn't tell me. That's why I was arrested. I'm a highly visible performance artist. They pick out members of the crowd they don't like, and later on, when you're taken to the police station, that's when they make up lots of charges.' The charges included resisting arrest, hindering police and assaulting police. Erwin, covering the protest, had our son Rainer with him, eleven years old at the time. Erwin gave him the office camera and Rainer crawled under a police car to photograph the Captain's far-from-gentle arrest.

Jim's own arrest came later, during one of the bravest acts of resistance that he's been involved in. Also one of the craziest:

> This feral mob from Melbourne had a firetruck, ran it on veggie oil. They had generators running on veggie oil, and they welded together this amazing structure and put it on top of their firetruck – a big revolving stage, made to look like a spaceship.
>
> This night a transvestite guy [no doubt Captain Starlight] was dancing on the stage, there were flashing lights, it was totally wild. They moved the whole thing onto the middle of the road ...
>
> The police came, looked at first like they were going to tip that firetruck over. They all got on one side of the truck, lifted it up and walked it off the road.
>
> Everyone sat down then ... It was very memorable. We all linked arms, sat on the road, maybe thirty to fifty people, guys dressed up

like people from *Star Wars*, mocking the special forces from Adelaide who were there for protests. They had batons, shields, helmets, all lined up across the road, in formation, I don't know how many, twenty, forty of these guys.

The sergeant said, 'Right, we're going to move you off the road. And batons will be used.' He said, 'Move!' and all the cops moved about a metre forward, stood there with their batons raised. He said, 'Move!' and they moved another metre forward.

To my amazement nobody got off the road that I know of, everybody sat there and linked arms. I was pretty scared, I'm sure everybody else was too.

Basically the cops just stopped. When they got up to us, they didn't hit us, they put all their batons down and started to drag us off the road, one at a time. Only those who kept going back ... were arrested. I did, most people didn't, I remember a cop saying, 'Grab him, he's been back on three times.'

Then I was taken to the watch-house – only three or four of us there.

Jim describes these people as feral, which in Alice Springs at least has come to have a very pejorative meaning, but he seems to like their spirit. He doesn't see them at actions much anymore, but they were around a lot from the 1980s to the early 2000s:

Sometimes they described themselves as ferals. They lived wild lives, often in the bush, did a lot of actions against logging, climbed up trees, lived in trees. They weren't so much into peace, love and brown rice, they were drinking a lot, big on alcohol. Their drug-taking was pretty hidden, but they smoked a lot of marijuana, that was obvious. They were not into nonviolence at all, not particularly aggro but they had no problems smashing stuff up, punching a cop back if they hit you. Their lifestyle was terrible, but they were very brave, they gave their lives to a cause. I admired that, there was a lot to admire about them.

~

Jim's next action at Pine Gap, and the court case that followed, would lay the foundation of his friendship with Margaret. This time, December 2005, almost two years into the Iraq War, there was just a small group involved. It was Jim's idea:

> I was pretty fired up about the invasion of Iraq. Maybe early 2004 I contacted everyone I knew who might be prepared to do a heavier action at Pine Gap, something arrest-able, that would get gaol time. It was not a big list. I wrote letters, telling them I wanted to do a Citizens' Inspection for Terrorism.
>
> This Terrorism Hotline had been started up by [Prime Minister John] Howard. I was ringing them regularly, saying that there's a terrorist base in Central Australia, involved in terrorist bombing campaigns, you've got to do something about them. Then they'd say, you got any names. I'd say, yeah, there's this guy, and I'd say the head of the base, whose name I knew at the time, I've forgotten now. Stuff like that. I was hoping they'd charge me with vexatious calls and I could go to court with that one but they never did.
>
> With the Citizens' Inspection, my intention was to sneak in, do an action, say we're trying to inspect the base for terrorist activity, see what happened. [Margaret's husband] Bryan Law was very inspired by the idea, so was Donna Mulhearn.
>
> Bryan took over a lot of aspects, which was good. He wrote to the minister and the head of the base to ask for permission first. That's when the minister threatened him with seven years' gaol for trespass under the [*Defence (Special Undertakings)*] Act if he did go in. Bryan told them we were coming, on such and such a date we were going to do the action. … I would never have notified the police or the authorities that I was going to do an action like that. My thought would have been, they're going to stop us. Worked out fantastic in the end, they couldn't stop us. Whether that was a miracle, brilliant planning, just good luck – it's up to everyone to decide.
>
> Donna said she wanted to come, Sean O'Reilly, Jessica Morrison and on the way we picked up Adele Goldie in Rockhampton. She was there for another court case, a Talisman Sabre action which had

> happened earlier that year. People had been arrested and had to go back for court. Adele was very keen to join us.
>
> So five of us made the trip in our old blue van to Alice Springs. Donna flew. We drove to Rockhampton, Townsville, Cairns, did some publicity on the way. Bryan was from Cairns, and Margaret. They organised media and actions up there. And we picked up Terry Spackman, Bryan's mate, who'd been in the Navy in England during World War II, he was quite old, he came along.

It was a 'bloody awful long hot road', Bryan Law recalled, coming into the Territory on the Barkly Highway, down through Tennant Creek and on to Alice. Bryan looms as a large presence in this action, the subsequent court case and still as a guiding spirit to the Pilgrims in 2016, despite his death three years earlier while awaiting trial for a rare Australian 'Ploughshares' action. In 2011 during the Talisman Sabre war games – a biennial training exercise for the Australian and US military – he had disabled a Tiger Attack helicopter at Rockhampton Airport.

A Ploughshares action, he wrote in his blog *Peace by Peace*, 'physically disables a weapons system in active use, in witness to the prophecy of Isaiah': 'They shall beat their swords into ploughshares, and their spears into pruning hooks: nation shall not lift up sword against nation, neither shall they learn war any more.' It would be more accurate, though, for the definition to include *intention* to disable. A hammer is the typical tool; it can be surprising how much damage a hammer does but it can also be relatively insignificant. Nonetheless the blows are central to the actions. Much of the property damage caused by peace activists in Australia is incidental to the focus of their actions, like cutting through fences or breaking padlocks, or else symbolic, like daubing blood, or focussed on conveying a message, through graffitied slogans. These actions may be carried out in the spirit of Isaiah, and share elements of Ploughshares actions, including the use of blood (often their own) and trespass on military installations, but they don't meet the definition offered by Bryan. His blog and other writings show someone grappling all the time with the ideas of nonviolent direct action (NVDA). Its most important principles were to 'disclose truth and accept consequences', he wrote. The Rocky Tiger Ploughshares action demonstrated both; the forthcoming trial he hoped would amplify them.

I had thought of the Ploughshares movement as historic. Not so. The Berrigan brothers were priests associated with the movement in America, its most prominent figures from its first action in 1980. They were often spoken of by my parents, who supported social justice movements in quieter ways, particularly through letter-writing, but admired utterly these brave men who spent years in gaol for their principled actions. When I was first talking with my mother, Patricia, about the Pilgrims and the kind of traditions they were working in, she pulled out an old clipping that she had tucked into a book. It was the *Sydney Morning Herald*'s obituary for Philip Berrigan in 2002, headed '"Turbulent priest" fought for peace'. The photograph shows him smiling, animated, handsome, tall. It takes a moment to become aware of the handcuffs and chains around his wrists. He had probably left the Josephites by then. He married Liz McAlister, formerly a nun, and they had three children, but he, and she too, continued to participate in Plowshares (American spelling) actions and to take the consequences, spending some eleven of their twenty-seven married years separated by prison. They tried to avoid doing actions that would send them to gaol at the same time, so one of them would be at home for the children. Daniel Berrigan, who remained a priest, served a total of nearly seven years in prison. He is dead now too, but the movement is alive.

It extends beyond the US and has drawn in some younger activists, including Australians – notably Ciaron O'Reilly, a founding member of the Catholic Worker community in Brisbane that Jim is a part of. He served thirteen months in a US prison for his part in disabling a B-52 bomber during the 1991 Gulf War (O'Reilly used his hammer on the runway tarmac). He was thirty at the time. In 2003 he joined four other Catholic Worker activists in Ireland in the Pitstop Ploughshares action, again disabling a US Navy plane. After two mistrials, the group was finally acquitted, three and a half years later. At the time of the action, O'Reilly was the oldest of the group, in his early forties; the two women were in their early thirties; the youngest was in his early twenties. However, it has been mostly older people who pursue this exacting form of resistance.

On 4 April 2018, seven men and women calling themselves the Kings Bay Plowshares broke into the Kings Bay Naval Base in St Marys, Georgia, intent on drawing attention to the threat to all life on earth presented by the world's deadliest nuclear arsenal – two dozen submarine-mounted Trident D5 missiles. A Jesuit priest and six Catholic Workers – among them Martha Hennessy, the 62-year-old granddaughter of Dorothy Day, and Liz McAlister, then 78 – the group speak with the fervour and broad aspirations for a better world that the Pilgrims share. The risks they faced in breaking into nuclear establishments were great, not only serious prison time – years rather than months – but potentially lethal responses. The military is authorised to use deadly force at these bases, and at times weapons have been drawn against these activists. In a 2012 action, for example, at the Y-12 National Security Complex in East Tennessee, a full-scale nuclear weapons production facility, guns were drawn against the three Transform Now Plowshares activists, including then 82-year-old Sister Megan Rice. After their unprecedented breach of a top-level installation, the only person to lose his job was the security guard who did not draw his weapon, having read the situation correctly as a protest by pacifists.

The Australian response to dissent in our military bases and justice system is not as extreme as in the US but it is hardening. Not that Bryan's 2011 Rocky Tiger Ploughshares action is particularly a case in point. You can watch a video of it on YouTube, part of it news footage that went to air that night. Bryan was a tall, portly man with a genial expression. He would have known that he cut a somewhat comic figure, riding a red tricycle, dressed in his Peace Preacher suit and a Bob Katter-style hat. You see him calmly take a boltcutter to the padlocked gates; Graeme Dunstan, the veteran peace activist who had driven him there, goes to help but it's already done. Bryan, who had had a triple-bypass operation less than three years earlier, rides onto the airfield, waving aside a man on a mobile phone, calling to him to get out. The helicopter is some distance away. Bryan coolly pedals over. No sign of anyone coming for him until he's quite close. A car approaches, a ute with a flashing light on top. Unhurriedly Bryan takes up a garden mattock and strikes the helicopter. The blow breaks the fuselage, doing enough damage to prevent the chopper from flying. A few men in khaki overalls finally run in and take charge of him. In voice-over at the end Bryan says: 'This act was to show how easy it is to intervene, in Australia, in the machinery of war.'

Bryan's involvement with Christians Against All Terrorism at Pine Gap in 2005 had led to him experimenting with small affinity groups 'carrying out inspirational NVDA'. His focus was on 'how to maximise the impact of actions while being economical in human and material costs'. He thought his Rocky Tiger action had a chance of showing people 'the extent and nature of the wars they presently consent to and enable'. This was because he had done actual damage to a war-fighting piece of equipment, in contrast to the action at Pine Gap where, even though they had sent the base into lockdown, the focus was on 'information for public discourse, not disabling weapons systems'.

Around the time of the Pine Gap action he had also become a Catholic, 'seeking to discern the path to peace, striving to enact Jesus' commandments (which seem key). Love God. Love one another. Love enemies. Holding onto that stuff, and figuring how to apply it, seems like a full-time vocation.' Margaret says he also saw that Catholics would 'take things a step further' with their social justice actions: 'Secular people were giving up, it was too hard, they weren't courageous enough.'

Bryan was already unwell in 2005. He was taking daily medication for diabetes, high blood pressure and a heart condition. During his interview with police, a few hours after his arrest, he referred to numbness in his extremities, especially in one thumb from being handcuffed, and his feet were 'sore and congested' from his long walk onto the base. Despite this and little sleep, he gave police a plucky, often funny account of the action:

> I wrote personally to Senator Robert Hill, the Defence Minister, asking for an inspection of the base. He refused. I wrote again reminding him of his obligations under the Nuremberg Principles to prevent the commission of war crimes, crimes against peace and crimes against humanity. I've had no reply. We gave him to the close of business on Wednesday the 7th of December to respond to our request and failing that we told him that we would undertake an inspection without his authorisation. ...
>
> When Senator Hill did not respond by close of business we prayed and examined our conscience and took an individual decision in relation to what we would do. Four of us decided that we were prepared to carry out civil disobedience and inspect the base. Two of us decided that they would play support roles.

Late on 8 December the four split into teams of two: Bryan would go with Donna Mulhearn, Jim with Adele Goldie. They would enter the base at different points while Sean O'Reilly (Ciaron's brother) and Jessica Morrison would be at the front gate to act as liaison after their arrest. (In this kind of nonviolent direct action, activists usually wait for arrest even if they're not initially detected; they don't seek to avoid the consequences of their conscientious law-breaking.) The four started walking their different routes at midnight. They were arrested in the early hours of the next day.

What was your plan of action? Constable Trevor Robertson asked Bryan.

'We prayed for a miracle.'

A miracle is what Donna felt they had achieved. She wrote a message to friends and supporters on 10 December, the day after their arrest: 'How on earth can a hotchpotch group of unfunded, untrained, unarmed (and in my case unfit!) Christian pacifists break into one of America and Australia's most significant military bases, and in relative ease, cause the base to be shut down for several hours? In a nutshell: it was a miracle! For heaven's sake we even told them we were coming!'

Jim was just as amazed: 'At 4 am Adele and I came close to the first three-metre high security fence. As we lay on the ground perhaps

500 metres from the fence, security vehicles drove nearby with their floodlight panning the area. We thought they knew we were there and were searching for us. At least twice we thought they must have seen us and our attempt to enter the base was over. Later we realised their surveillance was routine, and they had miraculously not seen us.'

After the patrols had passed he and Adele made the last dash through the open floodlit area. Adele hung their banner on the fence, a quote from Genesis as God condemns Cain for slaying his brother Abel: 'What have you done? Your brother's blood cries out to me from the earth.' Jim was carrying a barbed-wire crucifix, which Jessica had fashioned, and placed it against the fence before cutting the chainlink. They climbed through and he cut the second fence about ten metres in. Again they climbed through. The huge white radomes and satellite dishes were all around them.

Jim had fantasised about climbing one of the radomes but up close he realised that was not going to be possible. Adele headed for a tower next to a building and climbed onto the roof; he followed. They placed photos and leaflets on the roof and gave thanks to God. Shortly after, a security guard on a bicycle rode past and didn't see them, but around the back of the building he must have noticed the banner on the fence. Adele and Jim had taken photos of each other with a satellite dish behind them and radomes further back when the guard came into sight. He climbed the tower holding the dish: 'He must have looked around for a minute before seeing us. I waved, and he scrambled back down.'

Within a minute other security guards and federal police gathered below. One called out to them to come down. Jim told him that they had come to inspect the base for terrorist activity and would come down when they had something in writing from the commander saying they would not be stopped from doing so. A second guard yelled angrily that he was coming up to drag Jim off the roof: 'I responded that I would certainly not be surprised by violence as I was aware the base had been directly involved in the slaughter of thousands in terrorist attacks.'

Soon a number of guards and police were on the roof. The first one said, 'Get on your knees.' Jim replied, 'That's a good idea.' He knelt and prayed that the guard would withdraw his cooperation from the violence of Pine Gap, his prayer blocking out the further instructions that were being given. He soon found his head being pushed onto the metal roof with a knee holding it there.

The pair were taken down and driven to the front of the base, where they were searched numerous times before being taken to the Alice Springs watch-house. They were on a high. Their feat had left the base and its security arrangements looking stupid: 'We joked about how next time we should try it handcuffed and carrying tracking devices to give them a fairer chance of catching us!' And it wasn't over. Despite the repeated searches, they had still managed to smuggle out their photographs. On 10 December Donna posted online the now famous photo of Adele giving the peace sign from the roof.

The group believed Pine Gap was directly responsible for providing targeting information to the US-led bombing campaign in Iraq. They were able to quote strategic analyst Michael McKinley from the ANU on this. He said Pine Gap's contribution to the war was 'much more significant than any sending of Australian soldiers'. That made doing an action at Pine Gap very personal for Donna.

Ahead of the US attack on Iraq in March 2003 she had gone into the country, along with hundreds of other peace activists, to act as human shields at sites of critical infrastructure for the civilian population. She wrote to her supporters on 10 March, nine days before the US would begin its 'shock and awe' bombing:

> We moved out en-masse to the sites last Monday, so we now have all the human shields staying at five important sites around Baghdad. I'm based at the Taji food silo in North Baghdad. It stores and distributes wheat, rice and barley to 5 million people in Baghdad. It's a very important site for the Iraqis. I'm here with 10 other human shields and we plan to stay here throughout any military attack on Baghdad. We have notified our Governments and the US President of our locations, with the aim of deterring any bombing of this site.

She signed off, 'Your pilgrim, Donna'.

She shifted to a water treatment plant just before the attack started and stayed for the first shattering ten days of the campaign. In her memoir, *Ordinary Courage*, she writes viscerally of her experience of living under the bombs as well as her observations of the experience of ordinary Iraqis, the dead, the injured and maimed, including the women who miscarried their babies, and the children in a state of hysteria from the intense physical shock of the deafening explosions. She and most of the shields left when the ground invasion began, though a few remained to monitor the occupation and volunteer in hospitals. The sites they had been shielding survived the initial bombing campaign, though the first to be evacuated, the Al Mamoon Communication Centre, was destroyed within twenty-four hours. In December Donna returned to Baghdad to work with street kids, orphans and needy families, and was still there when the first anniversary of the invasion came around. 'The problem of achieving peace does not have a political solution,' she wrote.

> I don't believe that world leaders can bring us peace while ever they fight a war on terror using violence ... The solution is contained within each one of us. The solution is spiritual. It's found in forgiveness, reconciliation, compassion, reflection, tolerance, healing and love. It's not about religion, it's about you. Living in Baghdad with politically-induced violence all around me (huge explosion outside now) I'm convinced more than ever that this is what really makes a difference.

She quoted Gandhi, 'We must *be* the change we wish to see', and spoke about the inter-faith gathering she was organising for the anniversary, to be held at the National Theatre in Baghdad. Iraqi religious leaders would be there, Shia, Sunni, Chaldean, Sabian, Catholic, Orthodox: 'The politicians can argue about peace all day and night. We can just go and live it. Be it. Create it.' Six months later she was getting ready to come home again, knowing how lucky she was to be able to do that; she'd become so used to the sound of bombs going off at night that she'd roll over and go back to sleep; she'd been shot at by Americans, captured briefly and interrogated by mujahideen; she'd seen the best and worst of humanity, but her resolve was unshaken: 'The futility of violence is bleeding obvious.' This was why she had walked that night onto Pine Gap.

In the police interview room Bryan told Constable Robertson how the day before he had bought two boltcutters at Home Hardware in Alice Springs – 'a little bit of business for the local community'. He gave one set to Jim but was at pains to say not everyone had consented to cutting fences – Donna had not – and Bryan took full responsibility for that.

He laughed as he began the account of their walk through the night to reach the base, getting lost along the way. He declined to answer a question about whether they were driven to the starting point. They began walking at midnight, first following dirt roads for about six kilometres, eventually striking spinifex scrub at about 3 am – 'a very difficult walk'.

It was particularly hard for Bryan given his health problems, but Donna too had found it challenging, five hours trudging along in the dark. Her legs were aching and ankles sore, but 'It was worth every hole I fell into!' she told her friends and supporters after release.

Bryan admitted to cutting one fence on the privately owned land they went through to get to the base; said he'd be 'more than happy to pay' for its repair. Up close to the radomes – 'They wouldn't have been more than twenty, twenty-five metres away' – he had also started to cut through the inner boundary fence.

So, he was aware that he was on Commonwealth ground? asked Constable Robertson.

'Yes, I was aware I was on the Pine Gap Joint Defence Facility.'

Was he aware he was trespassing on Commonwealth ground?

'No, for me it's not a question of trespass, for me it's … it's a moral issue … I believe they have no moral legitimacy as a government because they took us into an illegal and immoral war against the people of Iraq, against the express wishes of the Australian people …'

So he believed he had permission to cut the fences, to enter Commonwealth land?

'No, they refused it. Senator Hill wrote back to us [as] Minister of

Defence basically threatening us with an offence carrying ... seven years' imprisonment if we went ahead.'

What he did believe was that they had a legal defence for their actions: 'In Queensland that would be the necessity defence ... taking action to prevent a crime being committed against property or person. In international law it would be a defence based on Nuremberg Principles. I'm afraid I'm not familiar with the Northern Territory jurisdiction ...'

Constable Robertson returned to what Bryan had actually done. He didn't ask about what he was wearing but Bryan brought it up anyway. He was in white overalls with a 'Citizens' Inspection Team' logo on them, embroidered in red. He was hardly trying to be inconspicuous:

> I've got to say it was amazing that the last two hundred metres of the walk ... this was a white set of overalls and we had security vehicles passing us all the time. We had one security vehicle stop behind us and point a spotlight in our direction ... and I waved to them and we were seen but not seen and nobody attempted to stop us or to apprehend us until I walked right up to the perimeter and pulled out my bolt cutters and started cutting and then they went 'Whoa hang on there' and then they got heavy with me.
>
> Officers of the Australian Federal Police put their hand on their firearm. Told me to, what was the phrase, 'Step away from the fence and get on your face on the ground now' and I asked them if they were going to shoot me and they just repeated that statement and ... and got very hostile and aggressive and I did for a moment figure that one of those officers might have pulled out his gun and shot me but I have to say that I noticed that no officer actually drew their firearm and I felt like, well you know, like I do not wish to participate in being shot at, so I dropped my bolt cutters and sat down on the ground and said 'Well look yeah okay, arrest me.'
>
> But I was made to lie face down on the ground in the dirt while officers put their knees in the small of my back, put my arms behind me, handcuffed me, rolled me over and I saw Ken ... the officer in charge of Commonwealth Police out at Pine Gap and I asked him

> for some dignified and respectful treatment and he just instructed his officers to follow their protocols and he left. He was clearly an unhappy man.

Bryan was carrying a folder with 'Christians Against All Terrorism' on the cover. Inside were what he called 'photographs of joy and hope', showing Joseph and Margaret, his 'beautiful son and lovely wife'. There were Christian icons and leaflets explaining the activists' position, 'what we understand to be the activities of Pine Gap': 'We weren't there for a trivial purpose [...] and we weren't there to cause trouble, we were there to follow the moral teachings of Jesus Christ.'

As the interview drew to a close, Constable Robertson asked if there was anything else he'd like to say. Bryan wasn't going to turn down the opportunity:

> Yes, I'd like to say that one of the things that our group did was to consult with Pat Hayes who I believe is the custodian ... traditional custodian with responsibility for that area from the Arrernte people.
>
> And Pat being a Aboriginal lady?
>
> An Aboriginal man.
>
> Aboriginal, oh Patrick, is it?
>
> Yeah.
>
> Patrick Hayes?
>
> Well Pat Hayes was how he was named to me.
>
> Okay.
>
> He's a senior Aboriginal gentleman with custodianship for that particular land. He gave us his permission to cross that land and said that the Australian Government and the US Government have never asked him if they could put Pine Gap there ... and in the current circumstances, with the illegitimate nature of the Australian Government, we took the word of the traditional custodian in preference to them.

That wasn't all. Bryan spoke more about the opportunity the Australian Government had had 'to come to an arrangement' with the activists: 'We were consistently rebuffed and refused and so we are left with the choice of civil disobedience. Civil disobedience is a traditional accepted means of bringing about lawful, peaceful political change and I adhere to it, I adhere to Jesus … and I am happy to take responsibility for this.'

But did he break the law? Constable Robertson wanted to know.

'What's moral is not always legal and what's immoral is not always illegal,' Bryan replied. 'I believe on this occasion I broke Australian law, a fairly minor Australian law, in pursuit of a moral purpose and I would encourage all citizens to consider doing it.'

Constable Robertson asked again: 'Okay, is there anything else you want to tell us?'

'Yeah, when you gonna join us?'

LAMENT

The Pilgrims' vision for their action in 2016 was of many peacemakers surrounding Pine Gap at dawn, converging on the base from several points. They would play music, light fires at the boundary, not attempt to go inside, but shine light into it, form a circle of grief around it, lamenting the dead of war – especially those dying by drone strike in Iraq and Afghanistan, Syria, Pakistan, Somalia, Yemen.

It's hard not to see the action's emphasis on lament as rooted, at least partly, in Bryan's not-so-distant death. Margaret credits much of the idea to their friend Graeme Dunstan. He had been Bryan's driver for the Rocky Tiger Ploughshares action and, after Bryan's death, had to face trial for the action alone. He too grieved for his friend, 'a man of huge spirit and courage'. Graeme was a staff cadet at Duntroon in the early '60s, but was turned off a career in the military by the Vietnam War. He became a peace activist and cultural entrepreneur, famously one of the co-directors of 'Australia's Woodstock', the 1973 Aquarius Festival at Nimbin. He pursues his activism as a spiritual journey, guided by a Vajrayana Buddhist practice, and his acts of witness in the public place are for him an artform expressed in flags, banners and lanterns, bringing colour and presence to events all over the country, especially up and down the east coast.

Graeme had taken part in the Pilgrims' thwarted first attempt to go onto the base but withdrew from the second: his old bones couldn't take it. His inspiration stayed with them, though. He had shown them a poem by Rainer Maria Rilke, the tenth of the *Duino Elegies*. Like Rilke's Angel, the Pilgrims would move in lament through the landscape, to 'the gilded noise, the flawed memorial', 'the streets of Grief-City'. The poet's words, written almost one hundred years ago, seemed to conjure the military base itself.

In the end the action became simpler, of necessity. Their ambitious plans didn't get traction with enough people.

On 28 September, just five of them answered the call to enter the base. Trespass had become important to their idea of pilgrimage, coupled with lament. It would complete their journey into the desert and against war, putting their 'bodies on the line', a phrase that Margaret uses often.

The last time Jim and Margaret had met up was in 2015, protesting against the Talisman Sabre war games. The games have taken place every two years since 2005, and peacemakers have carried out actions there, including blockades and trespass, since 2007. It's an important target because of the US presence and the number of soldiers involved: 'Over 25,000 people preparing their minds and hearts for war,' as Margaret puts it.

The protesters do what they can to disrupt the games. This includes trying to prevent the conduct of live-firing. There's no official detail around the extent of the live-firing but it is confirmed to take place and includes the firing of missiles. The activists have had some success – holding up the live-firing for a whole eight days in the 2009 games, by Bryan's account. A real if not enduring 'beating of swords into ploughshares'. But tangible results for them are not the only thing that counts.

In the 2009 Talisman Sabre, Margaret had been one of the trespassers known as the Bonhoeffer Four, together with Simon Moyle, Jarrod McKenna and Jessica Morrison (she had been with the Christians at Pine Gap in 2005). Starting out in pre-dawn darkness, they walked onto the Shoalwater Bay base in Central Queensland and released balloons – an action as gentle as would be the Pilgrims' lament at Pine Gap. Reflecting on what they'd achieved, challenging the perception of its futility, Simon, a Baptist minister from Melbourne, quoted Catholic social justice thinker Thomas Merton: 'you may have to face the fact that your work will be apparently worthless and even achieve no result at all, if not perhaps results opposite to what you expect … concentrate not on the results but on the value, the rightness, the truth of the work itself'. This kind of thinking sustains the activists' sense of purpose in the David and Goliath scenarios they involve themselves in. Jim holds dear a similar message from the Catholic Worker co-founder Dorothy Day: 'Don't worry about being effective. Just concentrate on being faithful to the truth.'

Not that they *want* to be ineffective. In the Shoalwater Bay action, taking the name of Dietrich Bonhoeffer, German theologian and Nazi resister, was a deliberate challenge to then Prime Minister Kevin Rudd. It was not yet three years since he, as an Opposition frontbencher, had published an essay on Bonhoeffer as 'the man I admire most in the history of the twentieth century'. Rudd had stressed the importance of Bonhoeffer's challenge to faith as purely a concern of the inner person and argued for a Christianity 'consistent with Bonhoeffer's critique', which always takes the side of 'the marginalised, the vulnerable and the oppressed'. He had contextualised this for 'George Bush's America and John Howard's Australia' – this was after the disaster of the invasion of Iraq and three years into its occupation. Rudd raised pointed questions about war for contemporary American and Australian Christians with their concern for the sanctity of all human life. Yet here they were, the Talisman Sabre exercises again underway, and the Labor government he was leading planning to spend $100 billion on new submarines, ships and combat aircraft. The Bonhoeffer Four wanted to remind 'Brother Rudd' of 'how out of keeping' this military spending was with his great admiration for Bonhoeffer's life and principles.

It was her companions in this action who introduced Margaret to Bonhoeffer. His idea of a 'costly grace' struck a chord with her. She saw it like this: take a risk, you might miss the hoped-for result – ending war – but through such endeavours you could encounter God in new ways. This happened for her when two American soldiers found the trespassing group. The war game, in which the soldiers were playing the role of insurgents, was put on hold while they all waited for the police to arrive, the soldiers standing guard. The protesters formed a circle and started to pray for the dead of war. One of the soldiers leaned forward to join in, offering the name of his friend who had been killed in Iraq. For Margaret this moment 'in the wilderness' was just as important as any broader political impact. It helped her recognise something about the soldiers' experience of war and what could be their grief. She had not had much to do with soldiers before; uncles on both sides of her family had fought in the world wars but she grew up without knowing them. From 2009 on Margaret dubbed the protesters' actions at Talisman Sabre the Shoalwater Wilderness Pilgrimages.

Two years later, during the 2011 Talisman Sabre actions, Graeme Dunstan came up with their graphic, a dove with outstretched wings rising from the words 'Peace Pilgrim', white on a green background. Some t-shirts were printed, people started wearing them: 'An identity either sticks or it doesn't,' says Margaret.

You can't see the t-shirts in the photograph taken the night of the 2016 Pine Gap trespass, as the five are all wearing jackets. Jim's arm rests across Margaret's shoulders. Both are smiling. They look like they could be the parents of the younger men, who are smiling too with the same clear steady gaze and strong sense of purpose. All five had travelled to Alice Springs from Queensland. Margaret drove in a convoy of two vehicles from Cairns, while Andy, Tim and Franz made the long trip with Jim, sharing the space in his old van with drums full of biodiesel.

Andy was intending to act on his own, to do something highly visible, creating an image directly linked to halting the operations of Pine Gap and that would draw media attention. He is a readily noticeable figure, with his lively blue eyes, wide expressive mouth, nose-ring, and red hair cut into a kind of relaxed Mohawk. Appearance, though, was not what he was thinking about; he wanted to do a lock-on. Before leaving Brisbane he'd bought himself a bicycle D lock. At $65, it was the most expensive single object he'd paid for in over five years. And he wouldn't be able to re-use it: if he was successful, he'd stay locked on until the police took to it with an angle grinder.

Once he got to town, after checking out and rejecting a couple of options, he decided he'd go for the undercarriage of one of the busses transporting staff onto the base. There was a choke point on the approach to the front gates where he might be able to halt the whole convoy. He planned the action for the morning of 28 September, preparing a media release and spending time practising the manoeuvre. He knew it would be hard to get under the bus quickly, before police saw him. The idea was that his friends would gather around with banners to give him cover.

An abandoned quarry just off Hatt Road had been set up as a camp for the Disarm Collective. They'd organised a bus to bring people for the convergence from Melbourne, with some also travelling independently. It was the more serious activist camp of the two. Twenty-six meetings had gone into planning their campaign for the five days, from 26 to 30 September. Their website presented a coherent case for protest, including links to credible resources like the Nautilus Institute for Security and Sustainability and the Bureau of Investigative Journalism. It also provided practical information – bring a swag, water, solid boots, warm clothes for the cold desert nights, torches, batteries, an inverter for your car allowing you to be electrically independent.

This was the same site Jim and family had stayed at in 2002, called Bush Camp then. Fourteen years later, some of the same people had returned to set up the Disarm Camp. Jacob Grech, for instance, whose Melbourne-based group in '02 was known as Ozpeace. But there was also a new generation of activists like Andy.

The question for activists is always what to do about their concerns. Andy knows that his planned lock-on may have seemed small but he sees power in going to a place with his own body and conscience and saying, 'We don't accept this and we want to resist it'.

When the Pilgrims arrived in Alice, the Disarm Camp had not yet been set up. So they went to what was being called the Healing Camp, set up on some small claypans known as the Ghan Pans, off the Stuart Highway before it turns towards Adelaide. There were some good people there, including Graeme Dunstan in his Peacebus, the old Mitsubishi van he travels around the country in.

He welcomed them with a cup of tea. Soon, though, the camp's 'new age' spirituality and self-focus started to get on Andy's nerves. He disliked the dope-smoking of many and what he saw as their awkward acts of cultural appropriation, their lack of awareness of their privilege. And he couldn't see what they were contributing to the convergence.

At the Disarm Camp he practised rolling under the bus from Melbourne and locking his neck to its axle. He is fit, not too tall, compact, he could do it, but he was nervous. Not out of fear – he knew police would become immediately aware he was under there, other protesters too – he just didn't want to fail. He hardly slept that night. Rain hammering on his tent didn't help.

Next morning outside the front gates, a small group of protesters huddled under umbrellas or in rain gear, all except for Andy. As the busses approached, Tim, Franz and Jim moved to block their path, holding up a banner and placards. The banner was Jim's favourite quote from Genesis about 'your brother's blood'; it had done long service at anti-war protests over the years.

Margaret placed herself right in front of the bus driver's window, ready to raise the alarm. Andy ran in and slipped under, wriggling on his back towards the front axle, getting wet through in the process. He managed to get the lock over the bar, put his neck in it and was trying to click it shut when he felt the hands grabbing hold of him. He held onto the axle with all his strength, but it was three to one. The police dragged him out, confiscated the lock, pushed his friends out of the way. The bus started to move before everyone realised it was over. Margaret was still crying out 'There's someone under the bus!' as she was lifted off her feet by a burly officer. It had all taken less than a minute.

The failure was hard for Andy. The adrenaline drained away; soon he was feeling tired, cold and deflated.

Entering the base that night became his plan B. Six people had tried on the first attempt; two, including Graeme, had since pulled out. Margaret wanted someone to take photos and live-stream video. Andy could do that. In any case, they were his friends – they would all be glad of one another's company.

After getting some rest, in the afternoon Andy, Jim and Franz went back to Hatt Road. A Disarm blockade fizzled when police did their own, stopping all traffic in and out. This forced base staff to use the back way, through private land off Ilparpa Road. The move also stranded Jim's van on the wrong side of the police cordon, so the three took what they needed from it and walked out, hitching a ride on the Stuart Highway. They were heading to a 'faith and activism' gathering at Campfire in the Heart, a retreat on the outskirts of town. This would help centre them ahead of the action. Andy recalled later:

> People around the group shared different perspectives, but of course what we didn't mention was the spiritual practice we were about to undertake – a pilgrimage into the eyes of Babylon, risking imprisonment to publicly state resistance to the US military rule of the world. 'Put away your sword,' Jesus had said, 'For he who lives by the sword shall die by the sword.' For me, faith and political action are indivisible. The pilgrimage we were about to leave on was a deeply spiritual act.

At the same time it also involved planning and a few logistics. Andy practised using Margaret's smart phone to live-stream. He's reasonably adept technically but doesn't own a smart phone. He knew that once they were arrested, which they fully expected to be, the phone would be taken from them. So live-streaming the action was essential.

The others were preoccupied with arranging space blankets into wearable form. They'd puzzled over how the police found them so quickly the first time. Was it by thermal detection? Someone came up with the idea of wearing space blankets to repel heat-seeking radar. That theory seemed improbable to Andy, stupid even, but what if they were caught and he was the only one not wearing the poncho they'd come up with? He'd blame himself and maybe everyone else would too.

So he relented. Margaret thought they looked hilarious, 'groovy dudes going to a disco'. The other novel preparation was that Jim put on shoes.

~

Margaret has her viola, her most precious possession, and Franz has his guitar. The two have composed a lament which they'll play once they reach the base. Margaret had put away her viola when she found music was getting in the way of her activism but about ten years ago she took it up again. Franz is studying music at university and loves playing with her, she plays so beautifully. They find music a powerful thing to do in their actions, in the way that it provokes emotion, bringing people together.

For as long as Franz can remember he has known about the base. He was only five when the family drove over for the big protests in 2002. These years later it is Pine Gap's escalating involvement in drone-strike assassinations that troubles him most, the impersonality of this new mode of killing. The lament is about trying to express the sadness he feels for its victims. It will be the first stage of the action, an evolving process. As they play and grieve they will consider the change that needs to come, and that will reveal to them their next steps.

Franz and Tim checked out the route in daylight before the first attempt and they're carrying a map. They take it in turns to lead. Tall and strong both of them, they inspire confidence but they have to concentrate, things look so different at night.

At first the scrub is thin, interspersed with clumps of knee-high buffel grass and the odd fallen tree. They all spread out, thinking to avoid the thermal radar. A couple of houses are visible in the distance. Dogs start to bark. The Pilgrims stop, then start again. At times they lose track of one another.

Andy barely slept last night, but new reserves of adrenaline are keeping him going, the excitement of sneaking through the desert on a mission for peace with a group of comrades. He enjoys their companionable silence and the meditative rhythm of the long walk. There's an uplifting beauty in the darkened landscape and the star-filled sky, although the crinkling sound from the space blanket starts to get distracting.

Tim finds the blanket incredibly annoying. He is used to walking through the bush at night, adjusting his senses to the dark as he moves along, his hearing compensating for the loss of vision. The blanket is not only noisy, it's covering other sounds – his friends' whispers, footfalls, possible sounds in the distance. And he's getting very hot. Even though every night in the desert is nice when you live in the city, the blanket is kind of ruining it for him.

An old farm road leads them through the first gap and they arrive at a cattle fence running parallel to the range. They avoid the gate, stepping through the fence a little to the west before joining the road again.

Tim tries to turn his mind to the people affected by war through the intelligence-gathering of Pine Gap. He thinks of himself as a farmer primarily. The people most affected by the drone strikes are just simple farming people, so he tries thinking about them on their farms but also finds himself wondering how the Pilgrims are going to get past the line of police surely waiting for them. Not that he imagines he can avoid arrest, or even wants to, but he hopes to reach his goal first. Then he realises he's feeling annoyed again, wondering if the wretched space blankets are actually working to mask their heat signatures.

The terrain has become less forgiving, the scrub thicker, and spinifex with its sharp spikes has replaced the buffel grass. In some places the ground is rocky and loose underfoot and there's the odd ditch. They unnecessarily climb a hill and are halfway up another when Franz goes ahead to scout the terrain. He realises they are well off track. They have to head back east until they meet the old farm road again.

No time for a nap tonight though they're all getting tired, none more so than Margaret. She'd been going to a personal trainer in Cairns to prepare for the walk, but this is the second time she's done it this week, she hasn't been sleeping well and tonight it's all catching up with her. She takes Andy's arm and sometimes she takes Tim's.

Tim wants to keep pushing ahead – they can't miss the base, they're going in the right general direction and it will be lit up like a Christmas tree.

Exhaustion is kicking in for Margaret, she doesn't think she can make it and sends Franz ahead.

They're close now, about five hundred metres from the outer

boundary, close enough for the light from the base to affect the brilliance of the stars. They think they may have been spotted and for a while hide under some bushes before making the final spurt.

Andy finds himself wondering about getting arrested and charged under the *Defence (Special Undertakings) Act* with its heavy maximum penalties but he smiles to himself: 'Who wouldn't risk a bit of gaol for their friends when we're all here together, doing what we believe in?'

It is just past 4 am when they cross the boundary, another cattle fence, four strands of wire, barbed at top and bottom. There are trespass notices attached every forty metres but it's dark; in court none of them will recall noticing the signs or anything much about the fence, they are intent on their goal. Mariner's Hill rises steeply in front of them. It's a little over one hundred metres inside the fence. From its crest they will be able to see the base, an important part of their act of witness. But they haven't got long, already they can hear car engines starting up in the distance, and voices closer by. They hurriedly confer: they'll press on, do what they came to do.

The Australian Federal Police are in charge of the base's security services. Officers from around the country have boosted numbers for Operation Fryatt, in place for the duration of the Close Pine Gap Convergence. In the control room Protective Service Sergeant Matt Gadsby is monitoring CCTV. He's worked for the AFP at the base for ten years. About an hour ago he detected 'irregular movement' beyond the northern boundary and immediately called his boss Ken Napier (the same Ken who had overseen Bryan's arrest in 2005), getting him out of bed at his home in town. Then Gadsby got a response team together and called the Northern Territory police. They were on duty outside the main gates, stationed there over the previous few days to deal with protests.

CCTV now shows clearly a group of five people carrying backpacks. They're moving quickly. Gadsby's biggest concern is that they might be creating a diversion. He gets two officers to check the perimeter while he keeps watching the five approach from the north. The camera, positioned on top of a hill, responds to motion and tracks the Pilgrims moving in. Once Gadsby sees them breach the boundary, he alerts the troops.

The Pilgrims are rushing now, but halfway up the hill Margaret says she can't go any further. She takes her viola from its case. Andy is behind her and whisper-calls to Franz to get his guitar. Margaret starts to bow, and Franz to strum. Miraculously the instruments are in tune. It won't be the settled performance they hoped for but, to Franz's surprise, the music gives Margaret the spur she needs.

They keep clambering up the slope. Tim is right alongside them, carrying an amp plugged into Franz's guitar. The music moves him. Lamenting is not a solution in the physical world, he knows, but to people who are alive it says 'what happens here is wrong', and to the people who have died as a result, 'you are released'.

For Andy the lament is also for our society's unthinking complicity in bombing people on the other side of the world. He's prepared to take risks for his strong views yet he doesn't see himself as a confrontational person – he doesn't like protests that shout at people, self-righteously. He appreciates the lament for that, for its compassionate spirit. These thoughts, though, come later. Right now he's focussed on trying to get the beam from his head-torch onto the moving group so that he can live-stream. He's feeling his lack of experience with the technology, and the terrain is not making it any easier. Nor is the fact of the police being on their heels.

Jim has taken a poster out of his pack. It shows a turbaned Iraqi man holding a little girl, possibly his daughter or granddaughter. Her face is blood-spattered, eyes closed. She is wearing a long fluffy mauve cardigan, bedraggled now, over green track pants. What is left of her legs and feet hangs in a mangled mess. The turbaned man seems to have picked her up from a pile of dead bodies. The photograph was taken in 2003 following a US strike in Basra at the start of the war. Jim has used the image before. He even had it printed on a t-shirt that he wore for certain actions, with the words 'Liberated' above the image and 'Iraq' below, and on the back of the t-shirt, an image of his own family, growing up in peaceful privileged Australia.

Now as he moves up the hill he holds the poster in front of him and prays, for the victims of war and for the people working at the base: 'You always pray for your enemies, you love your enemies.' This is his act of witness. People are always asking, what can one person do? For Jim to actually go to Pine Gap and say, 'This place is an abomination, it's taking part in war crimes', even if only very few people listen, it's a powerful thing. Powerful for him, to feel like he is speaking out, doing something about what is happening. And it inspires other people. To be doing this action with Franz tonight is proof of that. He loves how his son has turned out, a lovely gentle person living for higher values.

Tim leads his friends to a spot on the shoulder of the hill from where they can see the base, all except Andy who is still below them, filming. Under blazing lights, the sprawling domes and buildings seem hyper-real. It's like looking through a high definition lens, Tim thinks. He feels a quiet sense of triumph at busting the myth of the place, a forbidden thing that they aren't supposed to even look at and yet they are doing just that, in spite of all the obstacles.

They are very close to the inner compound, it's maybe a hundred metres away. For Jim the sight is as obscene as the first time he saw it in 1987, when he'd danced at the back gates, dressed up as a cockroach. Only now it's bigger and its reach is everywhere.

Margaret thinks it's like coming to Mordor – vast and grotesque in this desert place, sacred to its First People. But the Pilgrims are shining a light on it, they're lamenting, and the police are coming.

The police are all over the hill. Some have driven to the top, others are coming up from the bottom, assuming the intruders to be somewhere in the middle. Back in the control room Gadsby has lost sight of them in the rocky folds, but the police soon get a bearing from the music.

Inspector Ken Napier is at the top of the hill. He's carrying a portable floodlight and shines it down the steep slope. It lights up the Pilgrims, right in front of him. He calls out for them to stop.

Margaret and Franz keep playing, their eyes fixed on the radomes, the blazing light only strengthening the impression of evil and the sadness they feel. For Franz it's as if everything stands still and he realises he is crying. He has never been so sure of the rightness of their action as in this moment.

I see them on the hill, bowing and strumming, transfixed and prayerful, in these lines of Rilke's that had inspired them:

> Let not a single one of the cleanly-struck hammers of my heart
> deny me, through a slack, or a doubtful, or
> a broken string. Let my streaming face
> make me more radiant: let my secret weeping
> bear flower ...

Napier has worked at the base for twenty-five years, rising to the top of its security services and in charge of the AFP presence in town. He has dealt with a lot of protesters in his time, and he and Jim immediately recognise one another. In fact Jim calls out to him, 'Hello, Ken. You're very fit!'

Napier is fit. He hung up his rugby boots only last season and he's a keen mountain biker. He and Gadsby helped found the local mountain-biking club, shaping the network of trails through the sandy and rocky terrain west of town. But right now he is breathless from running up and down the hill; he can't answer. Jim chuckles, 'Not really.'

Andy keeps on live-streaming. Napier gets enough breath to call out to other officers to grab him. Matthew O'Neill is closest. There's a short chase as Andy tries to buy himself a few extra seconds. O'Neill takes Andy to the hard ground, falling with him. His partner drops on top, wrenching the phone from Andy's hand. They twist his arms back and tightly cable-tie his wrists together before dragging him to the top of the hill. 'Hardly the worst treatment you could expect from the police,' Andy wrote later, 'but I mention it because when I got to the top I saw my companions all sitting around. Evidently they had been allowed to walk to the top unimpeded and not had a hand laid on them!' Andy now gets his best view of the base. Most photographs he has seen have been taken from a distance, showing the whole complex – the cluster of radomes like giant spider eggs, the tightly laid out low-profile buildings, and at night, the floodlit perimeter. But up close like this he is more struck by how banal this face of modern warfare is, the satellite dishes, the antennas – 'the banality of evil, as the saying goes'.

When he's taken down to the road, the cops who grabbed him from under the bus that morning are there: 'You again, hey!' They all laugh. It's been a long day.

Margaret plays to the end, perched on a rock just below Napier. He recognises her too, from Bryan's drawn-out trial for the 2005 trespass. They are driven people, Napier knows, but utterly committed to nonviolence; he respects them, even likes them – outside the base. Coming inside is when they take it too far.

Right now he's worried for Margaret's safety. He reaches down, takes her viola and then puts out his hand again, helps her up over a last big rock before guiding her down to the road where she is taken to Federal Agent Peter Davey. An investigator with the AFP, Davey is normally based in Perth. His role in Operation Fryatt has been to liaise with the protesters. He has visited their camps, spoken with them 'to build rapport', as he sees it, asking about their plans, offering his help to deliver to authorities anything like petitions, letters – legitimate actions. That doesn't include what's happening tonight. He asks Margaret if she has a permit to be in the Pine Gap prohibited area (she hasn't), issues a field caution and arrests her. Sure enough, she is charged under the *Defence (Special Undertakings) Act*. Maximum penalty for the offence, seven years in gaol. She's prepared for this. The Act was strengthened after the Christians Against All Terrorism successfully appealed their convictions. It's watertight now.

A female officer is called to search her and then she is whisked away to the Alice Springs watch-house, separately from the men who, after a similar routine, are taken together. The paddy wagons used for transport have a prisoner cage on the back, open to the air. Margaret has sweated under her space blanket, making her clothing wet, and somehow Franz has ended up without his blanket and jacket. Tim and Andy do their best to share their coverings with him, but Margaret is alone. She and Franz are chilled to the bone by the time they get to the cells.

THE LAW

What the Pilgrims are facing now, prosecution under this harsh defence legislation, had its geopolitical roots in events before any of them, even Jim, was born.

On 3 October 1952 Britain brought the First Nuclear Age onto Australian soil on the Montebello Islands, off the Pilbara coast in Western Australia. If I'd heard of the islands before, the name had been lost to me. Like many, I had come to associate the British nuclear tests in this country with one place – Maralinga in South Australia. The significance of nearby Emu Field had also dimmed in my memory until it was jogged in recent years by the Pitjantjatjara artist Kunmanara Mumu Mike Williams. He made paintings protesting the infamous legacy of the tests, including Totem 1's toxic 'black mist' that spread out across the heart of the country.

My mother, Patricia, remembers the name Montebello clearly. A child of the '20s, by 1952 she had graduated with a degree in economics, married and become the mother of a little boy; five girls would follow, of whom I was the second-born. At the time of the third test on the islands, on 19 June 1956, I had not long passed my first birthday; Jim Dowling likewise. This was the biggest of the British test bombs in Australia, Mosaic G2, much more powerful than anticipated or disclosed, dispersing a radioactive cloud over the entire mainland. In contrast to our parents, those of us born in the '50s, '60s and '70s, the decades of the deployment and atmospheric testing of nuclear weapons, carry its radioactive legacy in our brains.

When Australia was asked about testing the bomb here, Britain warned that 'the area is not likely to be entirely free from contamination for about three years'. Eventually realised to be a gross underestimate, this issue was not even mentioned when the hastily drafted *Defence (Special Undertakings) Bill* was introduced in the Australian House of Representatives in June 1952. It proposed to declare the Montebello group a prohibited area for defence purposes. That the islands had been chosen for ground zero had been announced only in the preceding month. The public was curious and excited about the test, Minister for Defence Philip McBride told the parliament. Some of that could be ascribed to 'natural inquisitiveness but some is and will be nefarious'.

The legislation would 'incidentally' serve the purpose of closing the area to the inquisitive, protecting them from 'physical harm as a result of the experiment'. That the government was far more concerned with security was made clear by the severity of the penalties provided for: not only specific acts of sabotage, but unlawful entry could attract imprisonment for seven years. 'I make no apology for that,' said McBride, although he also drew the attention of the House to a provision which would come to have particular relevance for the Pilgrims sixty-four years later: any prosecution under the legislation would require the consent of the attorney-general. 'This, I suggest, will afford a safeguard against the measure being applied without due consideration.'

The ANZUS security pact had been signed in the previous year, and would eventually become central to Australian foreign and defence policy, but Britain was still very much the Mother Country. Australia's interests were assumed to coincide with 'British Commonwealth power and defensive security'. The legislation would thus apply to any similar undertakings 'whether they are for the defence of Australia alone, or whether, as well as being for our own defence, they are for the defence of another country with which we are associated in preparing to resist international aggression'.

Closing his speech, McBride said that it would not be possible 'at any stage to make a detailed statement' about what was happening on Montebello but he assured the House that 'in no way will the interests of Australia be overlooked'. Pine Gap had not yet been thought of, but the secrecy that has always blanketed the 'prohibited areas' and the bland assurances about what goes on in them were there from the start, as were nuclear geopolitics.

The Bill passed into law on 10 June. Britain's device, a plutonium bomb of the type dropped on Nagasaki, was duly exploded on 3 October in the hull of a ship off Trimouille Island in the Montebello group. Today the islands give their name to a marine park, promoted as an 'explosive attraction', a destination for 'dark tourism'. Visitors are advised to limit time spent at the old test sites to one hour per day, but in the years after the tests the sites were neither cleaned up nor policed; so much for the government's concerns for safety. Salvagers ignored warning signs and took away the huge quantity of contaminated scrap metal and other waste that had been left lying around the sand dunes.

The detonation put Britain into an exclusive league of three, the Soviet Union having tested its first atomic bomb in 1949. For Australia and its great and powerful friends, the Soviet test had been an alarming development in the feared quest by communism for world mastery.

The declaration of other prohibited areas, such as the uranium mine at Rum Jungle, followed passage of the Act, but it was not until 1967 that several lots around Pine Gap were pieced together and so declared. The base began operating three years later, a ground station for satellites that would intercept electronic signals passing into space from the Soviet Union. The CIA ran it from the start but this was not known even by the Australian prime minister and defence minister at the time. Euphemistically called the Joint Defence Space Research Facility, its use for signals intelligence was not admitted in the Australian parliament until 1973. Disclosure of the CIA's role did not come until 1975.

The base expanded enormously in its technology, functions and mission over the coming decades – as did resistance to it, particularly in the anti-nuclear campaigns of the 1980s – but the special legislation protecting it went untested. Use of the Act, against protesters who went onto the bases at Pine Gap and Nurrungar in South Australia, had been threatened before but had not eventuated; they were instead dealt with summarily, under the *Commonwealth Crimes Act 1914*, for simple trespass offences. The 9/11 terrorist attacks in New York in 2001 hardened the government line on such actions. The Christians Against All Terrorism crossed it when they entered the base four years later. For their 'clandestine' actions, 'striking at the heart of national security', a simple trespass charge would not suffice. Then attorney-general Philip Ruddock authorised their prosecution under the *Defence (Special Undertakings) Act.*

Ruddock, at the time of writing, is the mayor of Hornsby Shire in New South Wales, having retired from federal politics in 2016. I asked him why he thought it was necessary to use this legislation against nonviolent protesters, why prosecution under the *Commonwealth Crimes Act* would not have sufficed. He replied that he was reluctant to put his views without access to the documentation before him when the decision was made. He said there could also be 'security implications about which it would be inappropriate to comment', noting that 'in accordance with our Law the matter was dealt with as the Court thought appropriate'.

All four protesters were charged with unlawfully entering a prohibited area, and three of them with operating a camera in a prohibited area (only Bryan Law had not taken photos). Maximum penalties, imprisonment for seven years and two years respectively. The four were also charged under the *Commonwealth Crimes Act* with intentionally damaging property – the fences they cut through. Using the *Defence (Special Undertakings) Act* was a decision that the Christians said blurred the separation of political and judicial powers: having the attorney-general involved made the institution of proceedings very vulnerable to political bias. This, though, would not be the focus of their defence.

Their legal team was headed by Ron Merkel QC. A former judge of the Federal Court, he had purposefully returned to the Bar earlier that year to advocate on public interest issues, and Donna Mulhearn approached him for pro bono assistance. He agreed and together with Rowena Orr, they pinpointed an intriguing weakness in the *DSU Act*. Orr did the first pre-trial argument. This was long before she became a household name for her dauntless appearances during the 2018 royal commission into the banks. She 'did a real good job', Bryan would write in his laconic way.

On board too, from the time of arrest through to the Christians' eventual appeal, was Alice Springs lawyer Russell Goldflam. He acted variously as advisor, instructing solicitor, occasional junior counsel and legal aid officer, an experience he found 'both exhilarating and challenging'. While many others may have been in greater need of his assistance, it was wonderfully refreshing for him to deal with these clients, 'so unfailingly peaceful, honest, thoughtful, articulate, courteous, compassionate, conscientious – and sober'.

Orr argued that the four could not be charged under the *DSU Act* because the Crown could not prove that the base was a prohibited area at the time. To do so, the minister's declaration, under section 8 of the legislation, needed to be necessary for the defence of the Commonwealth – and for this there was no objective evidence. The minister's subjective assessment was not sufficient. The language of the section was plain and unambiguous, Orr argued, that objective fact must be present as a precondition. The section made no reference to any subjective test of satisfaction, in contrast to other sections that did. For the Crown to read into section 8 additional words about the minister's satisfaction would be 'impermissible'. An acquittal should be directed by the court.

For the Crown, Hilton Dembo countered that without such additional words section 8 would be unintelligible. 'Who else would make the decision?' he asked. 'It would be a nonsense to say that anyone else but the Minister would do so. These additional words are required.' He went on: 'The hardship the defendants will suffer is self-inflicted. They were warned not to trespass, they chose to do so. To impute invalidity to the statute would not see justice done.'

The warning he was referring to was more than just the trespass signs on the perimeter fence at Pine Gap and along Hatt Road. In the lead-up to their action, Bryan had written to then Defence

Minister Robert Hill, requesting permission to inspect Pine Gap for terrorist activity, but informing him that the group would do it anyway:

> In the months after sending that letter to Minister Hill we were investigated by ASIO, the AFP, and the intelligence branches of the Northern Territory Police in Darwin and Alice Springs. The Defence Security Authority opened a file on us, and a covert AFP agent was assigned to our public meeting on 6 December 2005 at the Arid Lands Environment Centre in Alice Springs. We were subsequently interviewed covertly by an ASIO agent. (Covertly: an ASIO agent pretending to be some other kind of official)
>
> One has to assume that all these agencies found our group to be nonviolent. Otherwise they each enjoy ample powers of preventive arrest, detention, control orders, etc to prevent any suspected attempt at an act of political violence.
>
> Minister Hill wrote back to me, warning us of heavy penalties under the *Defence (Special Undertakings) Act 1952* (*DSU*).

That warning stiffened when Justice Sally Thomas ruled against the Christians' challenge to the 'prohibited area' status of the base. She accepted the argument of the Crown that words could be read into the Act: 'To have it otherwise would result in a farcical situation where the Crown would have to call evidence on matters of national defence which presumably the defendants could seek to rebut.' These were not matters for the court to decide, but for the government of the day, she ruled.

The Christians, however, were not finished with their challenge. They made a request for extensive discovery of documents relating to the section 8 declaration: they wanted to see what had been prepared back in 1967 for the minister's consideration, including documents from the Department of Defence. They also asked to see the relevant documents for the section 6 declaration made at the same time, that Pine Gap is a 'special defence undertaking for the purposes of this Act'. Both declarations were published in the Commonwealth Gazette of 9 November 1967. Could those declarations still be operative all these years later and in such changed circumstances?

At the hearing of the application, both Merkel and Orr appeared for the Christians. The minister would have exceeded his powers if the section 8 declaration was of 'indefinite duration', they argued. The declaration was needed only for so long as it was necessary for the defence of Australia. And the minister's powers under section 6 were not the same as under section 8. Documents that recorded his decision and the grounds on which it was made would reveal whether or not he had misconceived those powers.

Further, if the facility, after so many years, was now less relevant to the *defence* of Australia, if it was being used rather for *aggression*, that would be significant, they argued. It would not be protected under the defence power and the Christians would be able to raise the defence that it was their 'honest and reasonable belief' that Pine Gap was *not* used for the defence of Australia. The best evidence to resolve these issues would be the agreements or arrangements that recorded the defence or other purposes for which the facility was being used at the time of the Christians' alleged offences.

To answer, Michael Maurice QC appeared for the Secretary of the Department of Defence; Dembo for the Crown adopted his submissions. Maurice argued that there was nothing in section 8 to indicate that a declaration lapses or ceases to have effect when the area concerned ceases to be required or used for a defence purpose.

Justice Thomas agreed with him. She also agreed that the only time that the minister needed to be satisfied that the prohibited area was necessary for the purposes of the defence of Australia was at the time of making the declaration: 'I further agree that the s 8 declaration continues in effect until such time as the government sees fit to revoke it, or it is found to be invalid. I also agree that there is no evidence that the base at Pine Gap has ceased to be used for the defence of Australia.'

That was what the application was all about, obtaining the evidence to resolve that issue, but she denied the discovery.

~

These issues, even in summary, may seem rather arcane but they go to the heart of the matter, for the Christians Against All Terrorism, for the Peace Pilgrims who would follow, for this book: what goes on at Pine Gap – the full story – and what does it require of us as citizens? By probing the base's protective legislation, the Christians and their legal team were attempting to shed some light on that. With Justice Thomas's pre-trial decisions they didn't get very far, but they should have been allowed to go further than they did. That was what the Northern Territory Court of Criminal Appeal found when it later overturned the Christians' convictions under the *DSU Act*.

Their eleven-day trial concluded on 14 June 2007, with convictions of the four on all counts. The jury had not found it easy; they had been visibly moved by the Christians' testimony and deliberated for close to five hours. Although the Crown called for custodial sentences with actual time served, Justice Thomas more reasonably handed down fines ranging from $450 to $1250. By August the Commonwealth Director of Public Prosecutions had lodged an appeal in the Northern Territory against the 'manifestly inadequate' sentences.

Donna and Adele, represented by Merkel QC, cross-appealed, with Bryan and Jim joining the action later. Their grounds alleged errors by Justice Thomas in her pre-trial rulings, including the refusal of their applications for discovery which in turn led to 'a denial of procedural fairness'.

The Court of Criminal Appeal agreed with them, ruling that Justice Thomas had erred in reading words into the Act. Section 8 did require an objective precondition (as Rowena Orr had argued). Thus the Christians should have been able to test the validity of the declaration and to do this they should have been allowed discovery of documents about the purpose of the facility. Justice Thomas had also erred in removing those issues from determination by the jury: the question of whether the base was indeed a prohibited area was a factual matter which should have been the jury's 'sole province'. There had been a miscarriage of justice and the convictions under the *DSU Act* could not stand.

The court also rejected the submission by the Crown that it order a retrial. A new trial would again likely be lengthy and again involve legal arguments about the purposes of the base and questions of public interest immunity. The offences were at the lower end of the scale of seriousness and the defendants had already served periods of custody, in lieu of paying fines, that were in part for the convictions they had successfully challenged (their convictions for property damage under the *Commonwealth Crimes Act* stood).

The four had all been arrested for not paying their fines soon after they arrived in Darwin ahead of the appeal. An important principle of civil disobedience is accepting the legal consequences of your actions. The four had decided against contributing 'to the Government's coffers' but were ready to serve prison time in lieu. Donna and Adele were released before the appeal began. Bryan had to serve ten days and only got out early on the second morning of the hearing. Jim, with thirteen days to serve, was still in custody. The appeal judges were not very happy about that, and their enquiries during the lunch break led to him being freed on bail that afternoon. He was able to leave the dock and join his friends in the public gallery. Once the acquittal was ordered there was no more time to serve.

Chief Justice Brian Martin, one of the three appeal judges, warned the defendants against seeing the court's decision as 'a vindication' of their protest. The decision meant that they should have been allowed to test the validity of the declaration, but it did not mean they would have been successful. 'Until a court finds otherwise,' he wrote, 'the declared area of the Facility remains a prohibited area and it remains an offence to enter the premises of the Facility without a permit.'

Still, the Christians and their legal team were all smiles on leaving the court, perhaps none more so than Ron Merkel QC. The acquittal was seen as, at the least, an embarrassment for the Commonwealth. My report in the *Alice Springs News*, though, overreached with this lead: 'The top secret American spy base at Pine Gap will have to make a full and frank disclosure about what it is doing, on the doorstep of Alice Springs, next time a government wants to prosecute trespassers on Pine Gap land under the *Defence (Special Undertakings) Act 1952*.' When the reasons for the court's decision were published, they made clear that the relevance of evidence about what Pine Gap does is one thing; getting access to the documents providing that evidence is another. The documents the Christians sought should have been produced. However, if the Crown had argued public interest immunity, as it undoubtedly would have, allowing inspection of them would have been for the trial judge to determine.

My report also quoted Reporters Without Borders, the international press freedom organisation: they welcomed the acquittal and suggested that it made 'further prosecution of the activists and – eventually – the media under the *Defence (Special Undertakings) Act* of 1952 highly unlikely'. That too was overly optimistic. Justice Thomas had been more on the money in her sentencing remarks: by using the *DSU Act* 'the Crown have in effect given a warning they may act somewhat differently to future offenders. In this way they have signalled a general deterrence to persons who may be contemplating such actions. These particular defendants may well have proceeded on the basis that the usual consequence of their actions would be a fine. That situation will not exist for either them or like-minded persons in the future.'

In national media, legal issues commentator Richard Ackland included the trial of the Christians, under the hitherto unused *DSU Act*, in his 'impressive list of civil liberties violations in our relaxed and comfortable land'. The case also attracted the attention of Frank Brennan, Jesuit priest, lawyer and academic, who has a particular interest in civil disobedience dating back to the repressive Joh Bjelke-Petersen years in Queensland. His 1983 book, *Too Much Order with Too Little Law*, carefully examined the question of when civil disobedience is justifiable. Writing twenty-five years later, he considered what the Christians had done at Pine Gap to be 'an act of civil disobedience – the deliberate breaking of a law in order to protest some other law or policy'.

> Civil disobedience can be justified when citizens have tried all lawful means to reverse the offending law or policy, when they do not threaten the health or safety of others, when they are prepared to pay just compensation for any property damage caused, and when

they are willing to pay the penalty justly imposed by any court. Civil disobedience can be an honourable means of political protest.

In Brennan's view, Philip Ruddock, as the attorney-general who authorised their prosecution, had been 'hoisted by his own petard'. Referring to the grounds of the Christians' successful appeal, Brennan wrote: 'If he had wanted these civilly disobedient protesters to go to jail, he would have needed to provide the judge and jury with details about the purposes of Pine Gap. Parliamentary committees, juries and the citizen's ultimate right to civil disobedience are necessary safeguards for liberty when government is tempted to use the legal sledgehammer to crack the nut of political dissent.'

~

One year after the Christians' successful appeal the government amended the Act to close off the vulnerability they had pinpointed. The declaration of the base as a prohibited area would be specifically written into the legislation, and any work within it would be a special defence undertaking. This would provide 'a firmer basis for any future prosecutions by removing the opportunity for argument about the validity of a declaration'. This was Joel Fitzgibbon, Minister for Defence in the Rudd Labor government, speaking to the second reading of the amendments Bill. A new section would also make clear the purposes of the Act and that a range of constitutional powers supported it, including the defence power and the external affairs power. 'These protections are essential,' said Fitzgibbon, 'to a facility of such sensitivity and importance to Australia's defence and external relations to deter mischief makers and those with more sinister intent.' He glossed the purpose of the base in the usual terms, leaving out half the story: 'The Joint Defence Facility at Pine Gap makes an important contribution to the security interests of both Australia and the United States of America, through the collection of intelligence by technical means and the provision of ballistic missile early warning information.'

One by one, other Labor men rose to support him, all of them trotting out the same half story and none of them examining the case for using this legislation, rather than the *Commonwealth Crimes Act*, against protesters as in the past. For the Coalition only Robert Baldwin responded, unreservedly endorsing the amendments and praising the base as 'an outstanding example of the level of cooperation that has been achieved in Australia's closest defence relationship'. He ended by baiting former anti-base campaigner and musician Peter Garrett, then a minister in the Rudd government, on his past opposition to the base: 'It would be interesting to know his thoughts on these amendments, which will, in effect, make it easier to prosecute his fellow protestors.'

In 1986 Garrett had been prominent in the campaign calling for closure of the base, but he had changed tack before entering the parliament. In 2004, as he campaigned for election, he told the ABC: 'I don't believe that Pine Gap should be closed. I'm fully prepared to accept the position that Labor has taken. There is no doubt about it, that it is the threat of terrorism and the intelligence that we can gather from terrorism that is now one of the primary and most important things that Australia, in terms of our national security, needs to consider.'

In 2009, if he was present in the House, Garrett ignored Baldwin's bait and did not speak to the amendments Bill. His colleagues attempted to rebut some of the concerns raised by the Christians and their supporters. Mark Butler picked up on the worry over Pine Gap being a nuclear target: 'Were an all-out nuclear war to have broken out during the Cold War it is very clear that Australia would have been a nuclear target with or without the joint facilities in place.' He didn't explain the point further. As for sovereignty issues, it was 'an endless debate'. Mike Kelly was puzzled as to why protesters 'believe the world would be better off without it': 'it would be like losing an eye in a world where we need all our senses operating at full capacity all the time.'

Mark Dreyfus was more bombastic: the activists were 'supposedly pacifists', 'there was both conceit and naivety in their actions', which were also 'supremely arrogant'. Their assertion that Pine Gap was not a prohibited area because it was not then being used for the defence of the Commonwealth was 'absurd'; this legal defence was 'mistaken', said this QC who would go on to become attorney-general under Gillard. That's not what the Court of Criminal Appeal had found. The 'mistake' had been in the lower court's ruling on the defence. The defence itself was untested; that eventuality was of course what the amendments were seeking to prevent.

The man representing the electorate in which Pine Gap sits, Warren Snowdon, was more muted: 'The importance of this facility I am sure is well understood by those on both sides of this House. The consequences of damage or disruption are grave both in terms of our defence and that of our principal ally, the United States. As such, it would adversely affect our external relations with that country.'

Snowdon, like Garrett, had had quite an about face. In October 1987, newly elected to the parliament, he had hosted three southern colleagues, including a future minister, Robert Tickner, on a visit to Alice Springs in support of the Anti-Bases Campaign Coalition. Activists – among them Jim Dowling – were gathering from around the country to protest the signing of a new lease agreement for Pine Gap. The three visiting parliamentarians put out a joint press release backing the campaign's call for an extensive public debate to culminate in a full public inquiry into the bases. They wanted party policy nationally to respond to the concern that the bases were being used offensively. Arms agreement verification was essential, they said, but would be more appropriately done by an international monitoring authority. They argued that long-term leases for the bases were unacceptable; rather the leases should be reviewed in the life of each parliament.

In 2002, as another major protest against the base was being planned, Snowdon demanded that parliamentarians be given 'a security briefing of a high enough order to understand competently the functions of Pine Gap … and how they might have changed over time and may change into the future. This can be done and should be done without undermining our national security. After all we are "responsible adults" who are charged with legislating on the affairs of the nation.'

Perhaps this demand met with some success, not that the public is any the wiser. In 2019, as Snowdon campaigned for re-election,

Erwin asked him, for the *Alice Springs News*, about his attitude to the role Pine Gap plays 'in clandestine operations in other countries'. Snowdon rejected that characterisation as 'frivolous' and 'unsubstantiated':

> There are things where I have been privileged to get an understanding. Because of the nature of the understanding, I am not at liberty to talk to other people about it. I am not in a position to make any comment beyond what is available from publicly available sources about Pine Gap's information and intelligence gathering operations.

In the Senate in 2009, the chorus of unanimity on the amendments Bill was smaller but equally on song. The sole exception came from Senator Scott Ludlam. A contribution from Senator David Johnston, a future Minister for Defence, was notable for this bald statement: 'Pine Gap is run by Raytheon.' He was right, Raytheon had had roles at the base since the early 1990s and was by then the prime contractor for the base's site management, operations – including signals intelligence – and maintenance, having taken over from Boeing in 2005. No reflection on this corporatisation of intelligence roles from Johnston. He was more concerned that protests at the base were 'generally antinuclear or anti United States in nature'.

Ludlam took exception to that, being himself one of the people who had demonstrated at Pine Gap in 2002: 'In my experience, these people are not anti the United States; they are anti war. They were there, we were there, to protest the coming bombardment and the loss of life of tens of thousands of Iraqi civilians, and they will be back whether or not this Bill is passed into law.'

Only Ludlam, of all the parliamentarians who spoke to the Bill, had anything to say about Pine Gap being used not only for defensive purposes but also for aggression, providing targeting information for US air and ground forces, 'very likely' used in the invasion of Iraq and continuing to be used in attacks on 'so-called insurgents'.

Only Ludlam argued that Australia is kept in the dark about Pine Gap's full functions; everyone else was satisfied by the 'full knowledge and concurrence' credo. He pointed to the way the 1999

Joint Standing Committee on Treaties had been treated. Their brief had been to consider the continuation of the agreement between Australia and the US on arrangements for the base. They asked to visit Pine Gap for an on-site, private briefing: refused. (As democratically elected representatives, they had found this particularly affronting, although defence and security expert Des Ball told them they would have learned little from a site visit; understanding required analysis, he said.) They asked for national interest analysis from the Department of Defence: it was dispensed with in 147 words, the usual gloss. They ended up supporting the continuation in principle but noted 'the limited evidence made available'. Nothing had changed in the intervening decade. With the base protected from parliamentary oversight, Ludlam said, informed democratic policymaking was not possible.

Only Ludlam questioned the necessity of the amendments Bill: 'adequate legislation, in particular the *Crimes Act 1914*, already exists to protect Pine Gap from trespass or from acts of aggression'; the Bill was 'putting a very old Cold War piece of legislation on life support'.

The 'life support' worked: the legislation was now impregnable as the Pilgrims would soon find out, although not before a setback for the Commonwealth.

ENDS AND MEANS

At the Alice Springs watch-house the Pilgrims were given blankets and put into individual cells. Margaret got a dry shirt. Between mugshots and fingerprinting they all tried to catch some sleep.

Jim had been to gaol many times for his civil disobedience offences. The last time was in Darwin in 2008 in lieu of paying the fine he got for the 2005 Pine Gap trespass. Margaret and Andy had also spent their share of hours and days in police lock-ups. From experience, Margaret knew it would be noisy, hard to sleep; she'd secreted earplugs in her bra and used them now. Andy's arrest this time was the roughest of the five, but in contrast to past experiences, he thought the police on the whole were pretty civil. They all thought so and when they got to court Margaret made a point of acknowledging their good treatment at the hands of the Northern Territory police.

Tim's quieter presence and younger years belie an inner toughness. His first arrest came in 2014, during an action with Jim and others at the military base on Swan Island near the heads of Port Phillip Bay, south of Melbourne. As 'ultra-secret' as Pine Gap, if not more so, the island base is a training facility for Australian Secret Intelligence Service agents and others doing clandestine work, including defence special forces and in particular the SAS (Special Air Service regiment). Its communications facility is a control point for special operations, by their nature covert.

A Christian peace convergence had been held at the entrance to the island annually since 2010. I had heard about what happened at the 2014 convergence before I ever met the Pilgrims, from the writer Barry Hill, a longtime good friend. He can see Swan Island and stretches of the beautiful bay from his home in Queenscliff. When I've visited, we have often walked past the bridge that connects the island to the foreshore. At night it is lit up and that sight from his bedroom always reminds him of the futility of the Iraq War, its waste. In the early hours of an April morning in 2007 he woke to flashing lights across the water – three SAS soldiers, on their way back to the base from the pub, had driven off the bridge and drowned. They had been on leave from extensive service in Iraq and Afghanistan, to do counterterrorism training on the island. At the Christian peace convergence seven years later came another layer of associations, also of damage to young men.

Barry had previously joined the Christians on the bridge for their dawn peace vigils. That year, though, he felt ambivalent about their overly simple anti-war messages. The rise of Islamic State had changed the possibilities for America, and thus Australia, of getting out of Iraq and Afghanistan. Barry, like many around the world, had applauded the courage and skill of soldiers rescuing Yazidi women and children trapped on Mount Sinjar, where they had fled after Islamic State attacked their villages. 'Just the kind of war heroes we want,' he would write.

Then he heard from his friends in the convergence what happened to four of them on the island: soldiers – SAS in plain clothes, one of them masked – arrested them, stripped them, dragged them along the ground, pulled hessian bags over their heads, stood on their heads, stood on their backs, threatened them with drowning and anal rape with a stick. 'Unspeakable heroes,' as Barry put it.

Tim was one of the four. Eight had swum out to the island, intending to conduct a 'citizens' inspection' of the facility. Allegations about what the soldiers did to them were widely reported (the other four were arrested without brutality by the Victorian police). A Defence inquiry was held and admitted this much: the soldiers had handcuffed the protesters with cable-ties, pulled bags over their heads, stripped or cut away their clothing, ostensibly to search them.

Greg Rolles, another of the four, said he told two soldiers advancing on him that he was a nonviolent protester and wouldn't be resisting; still they tackled him to the ground. Pulling a hessian sack over his head, one of them said, 'Welcome to the bag, motherfucker.' No-one who hears of this can avoid thinking of that infamous image from Abu Ghraib – the tortured Iraqi man, wired up and head bagged, which prompted Sarah Sentilles's reflections in *Draw Your Weapons.*

The soldiers rolled Greg onto his stomach, pulled his pants and underpants down, dragged him by his handcuffed wrists along the ground for some ten metres, his genitals exposed. They cuffed and bagged Sam Quinlan, then cut his pants away with a knife, from bottom to top. Tim recalled they were swearing, 'fucking cunts and motherfuckers and all the rest of it', and shouting the whole time – 'Do you know the seriousness of what you've done? Did you know that you've endangered the families of the SAS and anybody who works here? The women, the children, you've put them in danger.' This confused him at first but then he thought, 'No, you have done that all by yourself'; he may even have said that out loud. The soldiers clearly knew who they were dealing with, peace protesters, as they referred to a radio interview that Greg had given by phone from the island earlier that morning.

The Defence inquiry's heavily redacted report said the soldiers' actions 'were based on logical decision-making and were reasonable in the circumstances', which included 'the overall heightened security environment'. The use of 'intimidating language', the 'blindfolding and searching' of the protesters 'was done in a manner that did not cause any significant injury to them … using the minimum force necessary in the circumstances'.

The protesters thought otherwise. They sued, three out of the four; only Tim didn't. Four years later they settled out of court, disappointed there was no acknowledgment of wrongdoing or public apology, but confident that procedures had changed as a result of their action – in Australia, that is. 'We will keep working for change for our sisters and brothers in affected war theatres, and encourage you to do the same,' they wrote to their supporters.

Barry's poem about the incident was published in *The Australian* and he dealt with it again in an essay titled 'The Uses and Abuses of Humiliation'. The captors and captives were all young Australians, around the same age, he observed, yet their only relationship was in opposition to one another. The soldiers had treated the activists as the enemy, their intention was to humiliate, but he was struck also by a kind of humiliation the soldiers were exposed to. In the Middle East 'our forces had experienced a defeat that could not be named' and the soldiers, according to the activists, were 'in denial or ignorant or both'. They were 'to put it crudely, the mugs of history, which in so many ways most of us are'.

'In the mirroring waters of Swan Bay, there they were, each side suffering a form of humiliation that seems destined to repeat itself *unless it is transfigured.*' Achieving that is about sustaining 'a human respect for one's enemy'. In striving for peace, 'the ends demand consistency with our means'. Here he puts his finger on one of the challenges the Peace Pilgrims constantly grapple with.

Before his Swan Island arrest Tim had already booked a ticket home to New Zealand. He decided to plead guilty from there, which he could do by letter; he was fined $200, which he paid, and it was all over. He's critical of this expedience now: 'If I was doing it again, I wouldn't plead guilty and I wouldn't pay the fine.' Perhaps though, moving on quickly helped protect him from the trauma of the events. It was one of the scarier times in his life. He too had been stripped, dragged along the road, had his back walked on, had his naked body leered over with remarks that made him feel sick, but he was also able to detach himself to a degree: 'I remember becoming more afraid after the hessian sack went over my head but even so the whole thing felt like acting and in a weird way the angry aggressive violence of the SAS just seemed a pathetic act.'

At some stage the Victorian police had arrived on the scene. They didn't do or say anything to intervene, but Tim felt they were uncomfortable with what was happening. When the soldiers moved away, Tim asked one of the cops to help pull his pants back on and he obliged. Eventually the protesters were loaded into a police van, the cable-ties were cut, they took the sacks off their heads themselves and were processed as normal. 'This being my first arrest I didn't know what normal was, but I was pretty sure you didn't normally get assaulted, stripped naked, and dehumanised by the police.' He never saw the SAS men again.

Tim had grown up with a tribe of cousins in the bush, a life full of rough and tumble, so physically he wasn't hurt more than he was used to. Mentally and emotionally, what protected him, he thinks, was that this upbringing was also 'full of love and understanding and critical thinking mixed with an emphasis on learning to be compassionate towards others. By the end of the day, I was feeling pity for the two men who'd attacked us. How horrible had their lives been to make them into the monsters they showed themselves to be? What horrors had they seen and had done to them as part of their training? How deep did the ugliness and hurt go?' Having managed to come out of the experience relatively unscathed, grateful for the people around him who love him and will listen when he needs to talk, he felt prepared to handle what would come from this new arrest at Pine Gap.

Franz, though, didn't know what to expect from this first arrest. His childhood memories are peppered with the arrests of his parents, his dad in particular. One that stands out was when Jim painted his own blood on the walls of a Raytheon building (Jim described it as an 'exorcism'); Franz was there, 'most of us were' – meaning his mum and brothers and sisters. Another was when Jim attempted a citizen's arrest of federal politician and minister Peter Dutton for 'war crimes' in Iraq; Jim in turn was arrested by police, charged and eventually found guilty of a public nuisance offence. These experiences were part of what made Franz feel his family was different to most around them. Distinctly upsetting was the time when Jim was injured by police. He had gone to a debate on the merits of a national ID card, at which Dutton was one of the speakers. Jim stood up the back with a placard at his feet, reading 'Peter Dutton Supports Terrorism'. He was bundled from the auditorium by security guards; outside, police gave him a working over and arrested him. *Crikey*'s detailed account at the time published a photo showing Jim's bloodied face, which Jim said happened as a result of a police officer pushing his head onto the floor and jamming his knee into the back of his head. Other painful manoeuvres followed. Jim was acquitted of the charges against him (breach of peace and obstruction) but his complaint about police behaviour to the Crime and Misconduct Commission went nowhere. The commission accepted the police version of events: any force used was in direct response to Jim's resistance and lack of compliance and the arrest had been 'in good faith'.

Franz was still in primary school when Jim went through the earlier Pine Gap trial. It was a huge deal for their family, dragging on for years – the trespass was in December 2005, the acquittal not until February 2008. Twice the Dowling children had been bundled into the old blue van to make the big road trip to Alice: the first time in 2002, and the second time, in 2006, for the committal hearing in the trial. Looking back, Franz thinks of it as an adventure but with a shadow hanging over it. In the past week in Alice Springs and last night in particular, inside Pine Gap, what happened back then had come into sharp relief; he was starting to see the bigger picture behind this significant part of his childhood.

~

One by one the Pilgrims were taken to an interview room where Federal Agent Peter Davey and his offsider questioned them. Margaret was first. The interview started at 6 am. She'd had no more than an hour's sleep and yawned often. Despite her exhaustion she mustered the energy to break into Davey's formality.

Davey told her she had been arrested under section 9 of the *DSU Act*.

'As you know,' she interrupted, 'I'm very familiar with the Defence (Special Undertakings) Act. I sat in some of the trial for three weeks in 2007 of the only people who've ever been charged under that Act.'

Davey continued to follow protocols to the letter. Was she a holder of a permit under section 11 allowing entry onto the base?

No, she had no permission from the US military nor from the AFP, but 'there are other authorities on that land'.

His turn to interrupt: he needed to inform her of her rights.

And so it went. Yes, she is an Australian citizen, a fifth-generation Queenslander on both sides: 'We come straight out of the Frontier Wars.'

As soon as she could, Margaret returned to the question of authority: She was acting out of a different type of authority, one of justice, of God's creational authority that speaks strongly against war. Davey's authorities 'I don't really pay much attention to.'

Davey asked her to talk about, in her own words, her unlawful entry upon Pine Gap.

'Well, that would be a very legalistic way of looking at what we did,' she replied. 'We went to a mountain to see the facility … I wanted to get as close as I could to bear witness to war-making … to lament in response to evil war-making in that place … It is causing a wave of death and destruction across the world that will fall back on this land eventually.'

She declined to say anything about how she got to where she started walking, returned instead to her own themes: theoretically it is a joint defence facility but that is a farce, it is a US military base. She got as close to it as she could. They would have gone closer but 'it was clear you guys had some type of infra-red thing' looking down on them.

Davey was impassive, as he's no doubt trained to be. He was also well used to the activists' discourse. In less formal settings perhaps he'd smile, share a joke. In the interview room there's no rapport-building in sight; he played it entirely straight for all of Margaret's cheeky provocations. He continued with his questions and she persisted in answering at her own tangent: the base is on stolen land, it is bizarre, spooky, a symbol of denial by the Australian people.

Later in the interview she linked this to the denial of Australia's Frontier Wars. On Anzac Eve she was involved in a commemoration of the Frontier Wars dead where she lives, in Gimuy (Cairns): 'A lot of people got killed and no-one talks about it.' That denial is why they don't pay attention to things like Pine Gap.

When Davey asked about crossing barriers, she said she was not sure: it was night, difficult to see, their aim was simply to get to that mountain, to play a lament for the dead at a place that wages 'illegal and immoral wars'.

About the other Pilgrims she spoke only of their kindness: 'Those boys held my hand in very difficult circumstances.'

On how they planned their action, she again said she was not sure: 'But we all came to be together on that mountain to lament war.'

'I can't hardly speak, because I'm so tired,' she told Davey.

He didn't respond, instead asking if she is aware that Pine Gap is a prohibited area.

'There's certainly a lot of secrecy and a lot of denial,' she said. 'It's very hard to get close to, there are a lot of police around it stopping people getting close.'

Was she aware that she was inside the boundary?

She hadn't seen the leases, so she was not sure about that and it was not her concern. She wanted to look the base 'in the face and stop pretending it doesn't exist'.

~

'If it is such a core element of Australia's national security, what Pine Gap does not need is legislative protection; it needs perimeter patrol – especially when Christian pacifists have politely provided forewarning of their intention to nonviolently enter the facility to pray. If it is indeed such a sophisticated intelligence-gathering facility, the capacity to gather intelligence about its immediate environment should perhaps be enhanced.' Ludlam again surely had a point with this comment when he spoke in 2009 against the amendments to the *DSU Act*.

In 2016, five days after the arrest of Margaret and the others, after further protests across the weekend outside the base and in town, and a well-attended peace conference held in one of the town's hotels, Protective Service officers at the base must have had red faces when they discovered another breach of the boundary.

Sergeant Gadsby signs on for the day shift at 6.05 am. He does a handover before he sits down at the CCTV monitors in the control room. At 6.23 he spots a 'person of interest' wearing a red jacket, walking slowly back and forth in the dry bed of Laura Creek, which runs east–west through the base.

At about 1.5 kilometres from the northern boundary, this incursion is much deeper into the base than the five's. Gadsby calls the response team for a briefing outside the control room. Lloyd Adams is on the team, assigned to Operation Fryatt from his normal posting in Canberra, where he's part of the Close Personal Protection team for the prime minister. At Pine Gap he's been patrolling the perimeter, mostly inside, sometimes outside. He's pretty familiar with the terrain by now.

It takes the team only a few minutes to reach the intruder's location. They drive into the creek bed and Adams gets out. The red jacketed man is walking away from them. Adams runs towards him; his partner, just behind, calls out. The man turns, puts his hands out from his sides and without being asked, drops to his knees. The officers cuff him: they call it 'gaining subject control' but the man offers no resistance. They caution and frisk him, twice. The man tells Adams his name is Paul Christie and gives his address in Cairns. He answers 'no' to a question about whether he is carrying a recording device or mobile phone but declines to answer further questions.

Peter Davey has been off duty in his accommodation on the base. It doesn't take him long to arrive, together with officers of the Northern Territory police. Davey asks Paul if he has a permit to be there.

'No comment.'

Davey issues a field caution and arrests him for unlawful entry.

In the watch-house Paul had a nap, got a cup of tea, had a smoke. His interview with Davey started close to 11 am.

Davey invited Paul to speak in his own words about his 'unlawful entry upon the facility'. Like Margaret, Paul resisted Davey's framework from the outset: 'I understand you believe you've got lawful grounds to arrest me ...'

Davey continued with the interview protocols. He told Paul that his answers to questions would include non-verbal responses.

If it's not a verbal answer, it's not an answer, challenged Paul: 'Culturally everybody has got different body language ... the only way I can answer any of your questions is verbally.'

He gave his personal details including mobile phone number: 'You can reach me on it any time to discuss the war culture.' He earns an income as a youth worker, pays taxes. He laughed when asked about his relationship status and declined to answer a question about owning a vehicle: 'I don't believe it is relevant to the interview.'

Davey asked how he was feeling. Paul took stock: 'Sleepy, sore feet, a bit of nervous energy, otherwise I'm calm.'

Once again Davey asked him to speak in his own words about his alleged unlawful entry.

'I understand you allegedly found me – or officers of the federal police, wasn't it? – found me on some land considered by the Commonwealth and the federal police to be restricted to those not holding a permit.' He said he was invited by 'elders of the Arrernte tribe', given permission to walk around on their land, 'to pray and sing for their children and children all over the world'. He did that with their 'permission, blessing and support'.

He said 'no comment' to a question about holding a permit. On his intentions, he repeated what he had said about praying and singing, adding as the object of his prayers 'the mothers and fathers of Arrernte people, the mothers and fathers of all the land because they live in a war culture and it needs to end'.

Davey and his offsider tried to get details about the route Paul followed into the base.

People had picked him up on Larapinta Drive, they were like 'angels'. In the car he prayed and prayed until he felt the need to go walking. He asked to be let out. The sun was setting; he wouldn't be arrested until it rose.

He followed a dry creek bed which spoke to him, saying 'go this way'.

'The way opened up for me, I kept walking and walking, praying and singing ... Light came and took away all obstacles.'

~

Paul had been involved in the first foiled attempt with Margaret and the others. He first met Margaret and Bryan in 1994 during a campaign trying to prevent construction of the Skyrail cableway in the Barron Gorge National Park, part of Queensland's Wet Tropics World Heritage Area. Paul joined the bush camp supporting tree-sitter Manfred Stevens, who stayed in the tree for 208 days, almost seven months. 'At the time it was the longest tree-sit in Australia,' recalls Paul. 'He even got married in the tree!'

Paul was going in a different direction culturally from Margaret and Bryan – 'further away from society', including away from established religion – but his friendship with them and embrace of nonviolent direct action endured. During their child-raising years he and Margaret both avoided actions that risked arrest but once his youngest daughter was well into her teens, the time was right to get involved again. He knew Margaret had done actions during Talisman Sabre, with the Bonhoeffer Four in 2009, with Andy in 2011. In 2015 he walked in with her: 'We went with prayer and made contact with soldiers.' They were duly arrested, charged with trespass on a Commonwealth facility, spent the night in the Rockhampton watch-house, went to court the next morning, pleaded guilty, were fined and released. 'For me it was a good action. It broke the ice.' He'd feared going up against militarism, the potential severity of the charges, compared to opposing developments like Skyrail. Now he was prepared to face that threat. The Close Pine Gap Convergence had provided another opportunity 'to embarrass the state and show them I'm not afraid'.

After the aborted first walk into the base, he didn't want to risk going in on the same track again and asked around among sympathetic locals whether there were other ways in. He got a few suggestions; the one he chose involved a long walk, seven hours, but it got him there.

He had hesitated about going in alone but then resolved to do it. Following the arrest of the five, he said he teased the liaison officers like Davey: 'We haven't finished yet, there's more to do, things to see.'

Walking that night, he could see only five metres ahead: 'All I knew was I had to stay in the dry river bed.' If he felt hard ground beneath his feet, he knew he'd wandered off track and would find his way back. Visibility was even more difficult the closer he got to the brilliantly lit base. At times he got down on his hands and knees and eventually had a rest under what he thought was a clump of trees. When the sun came up he saw that it was in fact one huge tree with multiple trunks. He walked along close to the inner fence, picked some wildflowers; he could see people inside. Once he reached the northeastern corner of the fence, he turned around and walked back to the point where he was arrested.

~

Was he intent on entering the prohibited area? Davey asked him.

'I was intent on walking the land with permission of the Arrernte people, to sing and pray for peace.'

What was that area?

'We're living on stolen land, there's no particular area that isn't stolen.'

Paul stayed on message: Pine Gap is 'an illegal base run by the US defence force carrying out murderous activity against children of the world'. To inform the Australian people of this, he is involved 'peacefully and nonviolently with a collective of peacemakers from across Australia and the world'.

Davey asked him about his planning. Paul spoke of the map he was carrying that any member of the public can buy, and that anyone 'with a sensible brain' would carry if they were doing a walk like that.

From the map, was he able to establish where he was?

No. Last night he got lost, he didn't have a compass and was simply 'wandering the river beds'. He had used the map when he first arrived in Alice Springs to find Hatt Road and the Disarm Camp.

And where does Hatt Road go to?

It's 'common knowledge' it goes to 'the US spy installation known as Pine Gap'.

With others he had engaged in song and prayer 'in front of the gates of Hell'.

Did he notice the signs?

'There are signs everywhere, this society loves signs.' He could remember the sign at the front gates, saying 'Stop'. He couldn't remember seeing any houses, caravans, gates or fences.

But it would be safe to assume they would be there, to stop people walking through, wouldn't it?

'That's an assumption for you that I didn't make, I followed where my prayers were taking me ... When I'm praying and shaking my rattle I don't think, I just walk.'

If Paul struck Davey as remarkable in any way, he didn't show it. He asked whether he was a member of political organisations, peace groups or other issue-motivated groups. Paul didn't refer to the Peace Pilgrims despite wearing their t-shirt and having already mentioned being involved with a 'collective of peacemakers'. He said he was 'just one man, facing a machine of death'.

Davey got to the contents of Paul's backpack. What equipment was he carrying? Paul explained the purpose of some of its odd items: a Tibetan singing bowl 'to pay respects to the element of the Wind', a silver goblet, to honour 'the element of Water', the ceremonial rattle he had been carrying, 'to pay respects to ancestors'. There was also a red cockatoo feather, the bird being the symbol of the Close Pine Gap campaign; a kangaroo skin to sit on, a prayer shawl, and tobacco, lighter and papers were 'because I'm an idiot'.

What was the pocketknife for?

'As I said to the arresting officers, it's something any sensible person takes in case they need to cut a bandage.'

Asked to describe the location of his arrest, he spoke of the 'beautiful Alice Springs landscape, tarnished in a horrifying way by one of the biggest war criminal hotspots in Australia, called Pine Gap military installation'. It was 'the ugliest sight' he had ever seen.

At the end of the interview, he spoke of 'facing a machine of death, sitting in a room of brothers and I don't understand what is going on'.

Then he was told he could go.

At this point Paul had not been charged with anything and so he wasn't brought before a judge. That's presumably because Davey didn't yet know if he would be able to use the *Defence (Special Undertakings) Act* after the previous Thursday's embarrassing appearance in court for the Commonwealth, as they sought to start their prosecution of the first five Pilgrims.

~

In the watch-house Franz was exhausted and very nervous. Beyond the formalities of name, address and so on, he, like Jim, Andy and

Tim, had declined to participate in a police interview. They had all managed to quickly confer and agreed to refuse police bail so that they would be brought before the court straight away – that would be better than having to hang around in Alice past the weekend, waiting for a court appearance. This was the usual strategy of the old hands. They had no idea that they had, temporarily, outplayed the Commonwealth.

The Pilgrims assumed they would all be brought into court together, but just after 3 pm Franz was called first – a matter of alphabetic order. He was taken into court on his own. Walking up the stairs from the cells and into the dock he felt terrified, especially at the thought of having to speak to defend himself. In the end, he was asked just one question: 'Are you Franz Dowling?'

Through his tiredness, the rest was a bit of a blur, but Judge Daynor Trigg of the local court was keeping the Commonwealth on its toes. He took a short break to look at the legislation and came back with the killer question: Did the prosecutor – at this stage a police sergeant – have the consent to commence proceedings of the attorney-general, or of a person acting under his or her direction? (Trigg had no idea – he said so himself – who the attorney-general was.)

No, said the sergeant, but it was being sought. His Honour could remand the accused or bail him.

His Honour did not agree. Section 28 of the *DSU Act* outlined two steps. Police could arrest and charge a person with an offence under the Act without consent – as one might logically expect – but consent was required 'before a prosecution can be instituted'.

The sergeant tried to stand his ground – consent was being sought – but Trigg wasn't having it: the defendant should have been bailed to a later date until consent was obtained. That's when the sergeant revealed that all five had refused to sign any police bail paperwork.

'The trouble is the prosecution has now commenced and that is in the absence of consent. I'm not sure that can be remedied post-fact,' said Trigg.

The sergeant asked for a brief adjournment. When court resumed so did the to-and-fro, but now the Commonwealth was represented by counsel – James O'Brien, from the Territory's Office of the Director of Public Prosecutions.

'Okay, Your Honour, if there's no charges laid and there's not going to be any remand or bail because there'd be nothing to bail or remand on, there needs to be some charge there in place,' he confusedly proposed.

Trigg explained that it happened all the time: a person is charged and bailed by police, given a notice to appear at some date in the future while the prosecution prepares its file, then comes to court and the proceedings commence.

'It's Commonwealth legislation not designed to make sense!' he continued, to the delight of the Pilgrims' supporters in the public gallery. 'Whoever has drafted it obviously has little or no understanding about how the criminal justice system actually works. The section is a nonsense, but it's a nonsense we are forced to live with.' (Its hasty drafting all those years ago was showing.)

By now, Franz was being given some whispered explanations by Russell Goldflam who happened to be the Legal Aid duty lawyer that day. Russell, of course, was well-acquainted with the legislation after his experience with the earlier trial, but in Trigg's court that day he could scarcely believe what he was hearing. How had he, so familiar with the Act, missed this flaw in the proceedings? He was mortified but at the same time relishing the Commonwealth's embarrassment. This 1952 Act was a primitive form of anti-terrorist legislation. Applying it to peace activists was 'using a badly designed sledgehammer to crack a walnut'.

O'Brien tried one last tack, wondering whether there was an implied consent for the police to act under the attorney-general's direction.

'Where's that?' Trigg challenged. 'Where is the direction from the attorney-general in writing to the Northern Territory police to arrest and prosecute anybody who commits an offence under this legislation at Pine Gap?'

A rather dazed Franz was told that he would be free to go as soon as he was processed downstairs. There were whoops from the gallery as he was led away and Jim and Andy were called. Then someone bayed, like a dog at the moon. The often cranky Trigg must have been feeling pleased with himself – he tried some humour: 'Can we get the dog squad in? There's something feral out there.'

A little over an hour after it had begun, the five Pilgrims joined their supporters outside the court. Police stood guard at the entrance. 'Freedom, freedom!' chanted a woman. Franz was already strumming his guitar, Margaret alongside him was holding her viola. 'Give us the lament!' Graeme Dunstan urged. He had strung one of his colourful banners onto a bamboo framework under the Alice Springs Law Courts sign. The banner showed a silhouetted figure against a neon-coloured sunset, beating a sword into a ploughshare: 'Spears into pruning hooks, Swords into ploughshares, Study war no more', read the favourite slogan.

The Pilgrims lined up in front of it, and Margaret and Franz played. It was a slow, understated piece, wistfully sad. The mood of the small crowd quietened, a man and a woman danced slowly.

Jim spoke:

> We came here to resist war crimes of the United States and Australia, and the preparation for the ultimate war crime of nuclear annihilation, all of which are contributed to a great extent by Pine Gap. Pine Gap has been involved in all US wars since Vietnam and is responsible for the deaths of thousands, if not hundreds of thousands of people. What we are doing is trying to lament all those deaths, as Margaret has said, and also to resist the ongoing killing. We entered the base with those purposes.
>
> We fully expect the attorney-general to charge us under the Defence (Special Undertakings) Act. The police are obviously out to get us, they tried to refuse bail for a trespass charge. The police have been told to get us, I am sure.
>
> We hope to continue the resistance in the courts and hope to expose the war crimes of Pine Gap in the courts. I call on everyone to continue the resistance as well, as best you can, while there's time left.

Jim sees any violence, any war, as sins against the sanctity of life – moral crimes. Legal understanding of the term 'war crime' is far more elastic. Its long and complex history of trying to find a balance between military objectives and humanity would never satisfy Jim. Even in the stark instance of using nuclear weapons the International Court of Justice has equivocated: the majority opinion in a 1996 Advisory acknowledged the possible legitimacy of a nuclear strike in an 'extreme circumstance of self-defence, in which the very survival of a State would be at stake'. The majority did, however, reiterate the principle that the right to injure the enemy is not unlimited, and that key limitations on the use of nuclear weapons remain the principles of distinction (between combatants and civilians) and prohibition on the infliction of unnecessary suffering. The current legal position isn't the end of the story, of course – pressure from civil society can help it evolve. That's part of what keeps the Pilgrims and other campaigners like ICAN going.

Something else that Jim said outside the court confused me – that the police had tried to refuse bail when, according to what the sergeant had told the court, it was the Pilgrims who had refused bail. Two different procedures. The one Jim was referring to was court bail, not police bail. Before Judge Trigg made clear to the prosecution the error of their having instituted proceedings, the prosecution had told Russell Goldflam they would oppose bail (so Jim really should have been taking aim at the Commonwealth, not the police). 'This was extraordinary,' Russell told me later. 'Had the prosecution been successful, the defendants could have been held on remand in prison for over a year before getting to trial.' But the prosecution's resolve on this tactic must have fizzled.

~

The Pilgrims went home and a couple of peaceful months went by. Then, on 13 December, Attorney-General George Brandis authorised the institution of proceedings against them. I asked him, as I had his predecessor, why he thought it was necessary to use this legislation against nonviolent protesters, why prosecution under the *Commonwealth Crimes Act* would not have been sufficient. By this time he was Australian high commissioner in London, having quit federal politics in early 2018. He declined to answer my questions.

One by one the Pilgrims received their summons: first court appearance in Alice Springs, 21 February. Jim emailed Russell to let him know, hoping they might not have to travel from Queensland to appear in person: 'Of course I have not even asked if you are willing / available to help us once again. It would certainly be wonderful, and we would be very grateful, if you are.'

Russell was willing. Well-known for his advocacy against violence and for justice, he had also been a Pine Gap protester himself. He joined the Alice Springs Peace Group shortly after arriving in town in 1981, supported the 1983 Women's Peace Camp and took part in the big protests of 1987. There's a photograph of him from that year, wearing a Close Pine Gap t-shirt, one arm resting on the shoulders of Richard Tanter.

Like Richard, but on the other side of the country, Russell had been politicised by Vietnam. Studying history and politics at university in Perth, he discovered that the war wasn't 'merely unnecessary, unjust and unwinnable: it was also criminal and insane'. If he was conscripted, he wouldn't go; fortunately, a few months before he turned nineteen, newly elected Prime Minister Gough Whitlam abolished the draft. Richard, studying political science at Melbourne University, also escaped being put to the test: his number didn't come up. He went on to do a master's degree in New York, and then years of academic work and activism on the East Timorese resistance and Indonesia, before turning his attention to nuclear weapons. It was in this context that he became interested in Pine Gap and began to write about militarisation, from both theoretical and practical perspectives. His job, as he sees it, is to record what is happening and make it accessible. 'The truth is on our side long term,' he says of the peace movement. 'It doesn't matter if it looks not so good in the short term, we need information, reliable information.'

Russell's opposition to militarism had deep personal roots. His father's parents left Germany when Hitler came to power and eventually found refuge in Australia, arriving by ship in Perth in the late 1930s with their two children. Most of their extended family, left behind in Europe, were murdered in the Holocaust. Russell grew up in this 'shadow of annihilation, not that we talked about it much. But still.'

A job with the Institute for Aboriginal Development brought him to Alice Springs. Its director at the time was Yankunytjatjara man the late Yami Lester OAM. Working with him also shaped Russell's opposition to Pine Gap: 'Yami opened my eyes. He took me to places of haunting beauty on his country adjacent to the lands that had been so casually contaminated by the black Maralinga mist, which he told me had cost him his sight.'

So we come full circle, from the circumstances and law that paved the way for that abuse of country and its people, to their contemporary manifestation – the covert operations of Pine Gap and the use of that same law against the Pilgrims.

THE SACRED

With a Supreme Court trial hanging over their heads, some might have expected the Pilgrims to take a low profile but they did not, far from it. Their action and trial fell right in the middle of the centenary years of the First World War, with towns and cities around Australia holding events of remembrance, and the nation pouring millions into ceremonies at Gallipoli in Turkey and Villers-Bretonneux in France, where a new museum was built. There was peacework to do.

In Brisbane Andy was looking for ways to specifically respond to all the war memorialising. The centenary of the first referendum on conscription for overseas military service was coming up. The campaign and the vote, which took place on 28 October 1916, with a majority rejecting conscription, had been a huge event in Australia. Why wasn't it being remembered too? He decided give an account of it in his blog, eventually compiling 'A brief history of Australian war resistance'.

Before the First World War Australia had compulsory military service (for boys and men) for home defence. It was alone among the Commonwealth countries to do so until Britain introduced conscription of single men in January 1916, extending it to married men in April. New Zealand and Canada followed suit. Even for peacetime military service, many in Australia refused to register – a fact that might surprise many Australians now. This refusal resulted in 34,000 prosecutions between 1911 and 1915, with 7000 terms of imprisonment imposed. So perhaps it should not have come as a shock when the No case in the referendum won.

The campaign had been divisive, with each side trying to convince the electorate 'that the blood of Australia's men stained their opponents' hands'. For his promotion of the Yes case, Prime Minister Billy Hughes was expelled from the Labor Party, and joined with the opposition to form the Nationalist or 'Win the War' Party. They were victorious in the general election of May 1917 and later that year, after the Allies had suffered a series of heavy defeats, he again put the conscription question to a referendum. It was again defeated, this time by a slightly larger margin.

On the centenary date of the first referendum – a day to the month after the Pilgrims' Pine Gap action – Andy organised an event to commemorate war resistance, putting out the word to come together at the Temple of Peace in Toowong Cemetery in Brisbane.

The heritage-listed cemetery is on a broad grass-covered hillside dotted with tall trees. Today it is bounded by busy roads but once you enter the grounds it still has the atmosphere of tranquil melancholy that many old graveyards have. In use since the late nineteenth century, it holds many graves of historic significance as well as many of returned soldiers and monuments memorialising particular conflicts and branches of the armed forces.

Directly in front of the main entrance is a large and immaculately maintained war memorial, unveiled on Anzac Day 1924. Built with money raised from the sale of Anzac Day badges, it is made up of the Cross of Sacrifice, with its clear symbolic association with Christianity, and the more secular Stone of Remembrance. Modelled after memorials in Commonwealth cemeteries across the world, they were the first of their kind in Australia. Some three thousand people attended the 1924 unveiling. So many wreaths were laid that the Stone of Remembrance was completely covered.

That same year, on 6 December, just a short distance from the war memorial, the Temple of Peace was dedicated. This was the site Andy chose for his gathering. Unlike the war memorial, it is not immaculately maintained although steps have been taken to protect it from vandalism. Historian of Australian war memorials and the Anzac tradition Ken Inglis recognised the rarity of the temple's anti-war purpose. Other memorials given the name 'peace', not 'war', usually commemorate the Peace Treaty of 1919 rather than express pacifist convictions.

The temple is really a mausoleum, containing the ashes and remains of loved ones of the man who built it, Richard Ramo, as well as eventually his own. Its structure is elaborate – dusty pink walls studded with floral emblems and garlands, stained-glass windows, a stepped canopy topped by a temple in miniature. The inscriptions on marble plaques tell a poignant story of personal loss – 'All my hope lies buried here' – and fervent anti-war sentiment: 'Cease! Ye Nations, your whirling Dance of Death, For your Creator has no desire to destroy your Soul and Breath ... Your Passions, your Greed will be your Doom ... and bring to million hearts the greatest bitterness.' 'Workers of the World be Brothers,' Ramo appealed. 'Mothers of all nations ... save thy sons and daughters from the God of war.' And when 'the red flag of humanity flies over the world, I shall not have lived in vain'.

It's not hard to see the appeal of all this for Andy, the passionate mixture of unconventional faith and politics. In the wake of the heavy losses of the Great War the temple's dedication drew a large crowd, 'thousands' according to Ramo's plaque, in an occasion organised by the Australian Rationalist Association, whose president, while honouring the dead, called for international fraternity and rejected war as an evil of modern capitalism.

Andy's event was tiny in comparison but it had consequences that he couldn't have predicted. While they all stood in front of the Temple of Peace, remembering the largely overlooked stories of Australian resistance to militarism, Jim's eye was drawn to the Cross of Sacrifice at the nearby war memorial, its sword fixed where normally the body of the crucified Jesus would be represented. This outraged Jim's sense of the sacred, and he resolved to remove it.

When he shared his plans with his friends, Tim immediately came on board, but Andy was worried. He appreciated the way his friends felt, that the sword was a perversion of the message of the cross; he liked their ideas for the action as an act of repentance for the wars and violence that Christianity had been used to justify over two millennia; but he feared a strong backlash. Interfering with the war memorial would likely be seen as a heresy, given the elevation of Anzac Day and its symbols to religious-like status; worse, despite its worthy intention, it could be seen as disrespecting the war dead – especially as the memorial is inside a cemetery.

Jim and Tim went ahead and planned their repentance ceremony for Ash Wednesday, the first day of Lent, which fell on 1 March. They would act openly with prayer, reflection and song, as is their practice; then they would wait for the police to come and arrest them. Their impending trial for the Pine Gap action was not something they allowed to trouble them.

Andy, torn between loyalty and apprehension, offered support but did not want to get arrested for this action; he didn't want to have to defend the affront he could see it would provoke. Taking it up to the military, to big corporations, is one thing; confronting other people's sense of the sacred, even if he didn't agree with its construction, was another.

On the day they gathered at Dorothy Day House and put together what they would need: ladder, pinch bar, hammer and anvil, bread and wine; Franz took his guitar; Andy, a camera. He would limit his role to filming and photographing the action and sending out information afterwards.

At the cemetery Jim set to work on the sword, levering off the blade, which was separate from the more securely fastened hilt. Then it was Tim, with hammer and anvil, who turned 'swords into ploughshares', or rather, the blade into something resembling a garden hoe. Franz played and sang, others read passages from the Bible. No police came, no caretaker; a sole jogger went by without reaction. A priest was present and said a short Mass. 'We shared the bread and wine of communion,' Andy wrote later, 'and, after all the rushing around, stopped to reflect for a while. It was very moving in the quiet of the cemetery; in the company of people who, though from different backgrounds, were united in our belief in the transformative power of the message of Jesus; and in our desire to live our lives accordingly.'

Still no police and quite some time had passed, so Jim and Tim decided to leave a note, giving their names and phone numbers and explaining what they had done with the sword and why. Back at Greenslopes, Andy sent out the press release, which quoted Jim, but provided his own phone number as the contact for the photographs and video. He also posted on social media. He had done as much many times before for actions in which other people were arrested and although he had qualms about this action, he wasn't ashamed of having supported it.

Next morning, police came to the house and asked him to come down to the station 'just for a chat'. He knew what was coming. He was worried about the legal consequences – given that he was facing a possible gaol sentence for the Pine Gap protest – and about the social backlash. At the station he was arrested and put into a cell briefly with Jim: 'He had been thinking about it for weeks and was at peace. I was a bit of a wreck.'

News media had picked up Andy's press release. When he and Jim were transferred to the Roma Street watch-house, TV cameras were waiting. Inside the pair were charged – wilful damage to cemeteries, maximum penalty seven years' imprisonment, the same maximum as the Pine Gap trespass. Then they were bailed. Outside Jim responded to questions: 'I believe we redeemed the war memorial, the desecration is the sword on the cross, I don't believe we desecrated anything.'

The first reports confirmed Andy's fears: 'Religious fanatics vandalise war memorial' ran the headline in the *Brisbane Times*, followed by 'Crusading cowards charged with smashing up a Brisbane War Memorial' on *10 News First*. With Jim and Andy in the lock-up, Jim's wife, Anne Rampa, had responded to questions. She had witnessed the action. What was the difference, she was asked, between this action and the actions of the Taliban in Afghanistan, where the ancient Buddhas of Bamiyan were destroyed, and Islamic

State's more recent destruction of 'blasphemous' artefacts in Palmyra, Syria? 'We're not attacking another religion,' she replied. 'We're trying to bring our own religion back to its proper and rightful position in terms of the violence of war. We are very shocked by the presence of the sword on the cross, which is completely antithetical to what Jesus said.'

Comments by Jim, from the press release, also got a run:

> Jesus's last words to his disciples before he was taken away were 'put away your sword' ... For the first three centuries, his followers largely obeyed these words and refused to kill their enemies. With the conversion of Emperor Constantine all this changed, and ever since Christians have blessed countless wars, and even lead [sic] their own. We come here today to repent all wars blessed by the Christian churches. We come to remove the sword from the cross on which our saviour was crucified. We come to beat it into a 'ploughshare', in this case a garden hoe. The ploughshare is a symbol of life. The sword is a symbol of death. We choose life.

Of course the *Brisbane Times* also sought reaction from authorities. Brisbane lord mayor Graham Quirk was 'appalled' by the 'vandalism', seeing it, as Andy had predicted, as disrespectful 'to all fallen soldiers and returned service men and women'. Comments by RSL Queensland state president Stewart Cameron were more trenchant: for the sake of 'a cheap publicity stunt' it defiled 'one of Brisbane's most sacred war memorials and sites ... that pays respect to the brave men and women who have served their country'; those responsible were 'the lowest of the low'. A spokesperson for the Catholic Archdiocese of Brisbane distanced the church from the Catholic Workers as a 'group of lay people' with no ties to the Archdiocese. It supported 'any police investigation into this matter. The cemeteries of Brisbane mean a lot to the families and friends of the deceased. These sites should be treated with the utmost respect.'

Andy didn't care much about the opinion of the church hierarchy, but he felt for Jim, as a devout Catholic, to be tossed aside by his church in this way. Though not surprised, as the church had long 'betrayed the nonviolent message of Jesus', Jim was saddened by it, especially as the bishop had refused to talk to him about what they had done and why. And Jim rejected the association of their action with damage to cemeteries as the monument is free-standing and not part of any graves.

The activists got more recognition from Greens councillor, Jonathan Sri. On television news he acknowledged that a lot of people would find the action controversial but, he said, the Catholic Workers 'do a lot of good work in the community as well. They practise what they preach.' On the whole, the tone of this report was more mystified than aggressive; it was the first of many that drew attention to the Catholic Workers' bare feet, in voice-over and images, but it described them as 'Christian activists', not fanatics, and gave Jim two goes at explaining the point of what he'd done and his preparedness to face the consequences.

Tim had gone from the cemetery to work and wasn't arrested until the next day. A week later the police came for Franz as well, to everyone's surprise. The video that Andy shot showed Franz singing and playing the guitar – his only role. Andy worried for him: 'At least I was used to the experience and had played a fairly active role.' He was buoyed, though, as friends and supporters rallied around the group, including thirty Catholic Workers in New Zealand who sent an affirming card, and Catholic Worker peace activist Ciaron O'Reilly who gathered signatures for a letter of support, including from Nobel Peace Prize winner from Northern Ireland Mairead Maguire.

Andy couldn't bring himself to look at the comments on his Facebook account, even when he wrote about the saga in his blog almost a year later, but they are interesting for what they reflect about the elevation to the sacred, for many, of soldiers and death on the battlefield.

The majority were negative, some very hostile, and these were the comments that attracted lots of 'likes'. One: 'You're a terrible person. Not all people who fight in wars are war mongers. This is a disgusting act of disrespect and vandalism dressed up as religion. Shame on you.' Another: 'Desecration not vandalism'. Another again: 'This act shows hate towards those Heros [sic] who gave Their life. If you agree with it you are a grub.' This woman continued to post 'grub' and worse, like 'burn in hell', in reply to anyone who posted in support. She also scoffed at the religious motivation behind the action: 'Have you physically seen this so called Jesus? Or do you just follow a fictional book of bed time stories? Is that the hero you believe in? From a book? My heros are ones who I know and those who I've known and sacrificed their life for me and for the safety of others.'

Andy might have been more troubled by posts that were critical while also anti-war. One: 'To thousands of families it ... is an assault on the living memory (mythos) of their loved ones who they believe sacrificed their lives (hence the cross) for the benefit of all. I abhor war, but this is insensitive and dogmatic.' Another: 'I think i have turned away from thinking actions such as these and plough shares are a good way to go about things. They seem to achieve very little. This coming from someone who was a very passionate Christian anarchist for years and is still strongly anti war & for peace making.'

~

From up in Cairns, Margaret tried to be supportive, responding to the negative social media with her own comments and challenges. Meanwhile, she and Paul were focussed on another way to respond to the war memorialising, by organising a Frontier Wars remembrance ceremony. It is all connected for them, the war memorialising, the cultural amnesia around the Frontier Wars, the blind acceptance of the American alliance and its military bases on Australian soil. Graeme Dunstan had been organising Frontier Wars remembrance in Canberra since 2011 together with the Aboriginal Tent Embassy. Drawing some of their inspiration from him, Margaret and Paul reached out to the local First Nations people.

Margaret and Bryan had had a long, though intermittent relationship with Gudju Gudju, an elder of the Gimuy Walubara Yidindji. They had held a ceremony for Sorry Day in May 1998 at which Gudju Gudju's father had spoken. More than a decade later, on Anzac eve 2011, Bryan had organised a Frontier Wars lantern vigil in which Gudju Gudju was involved, on the esplanade in Cairns.

As the centenary commemorations of the First World War began in 2014, Margaret and Paul, with their friend and fellow activist Trish Cahill, organised another. By 2016 Margaret had also met Murrumu Walubara Yidindji and for the ceremony that year she and Paul worked closely with him.

Murrumu had spent time in Alice Springs – 'Mparntwe, yeah?' – as a reporter with the ABC; later he went on to work in the national press gallery. In those days he went under the name Jeremy Geia. Since 2014 he has renounced being part of the Australian state and lives outside its economy – that is, without direct access to money – and outside of its regulation regarding things like proof of identity and licencing. He's spent time in gaol as a result. He and other Yidindji have asserted their sovereignty and formed the Sovereign Yidindji Government; Gudju Gudju is the chief minister, Murrumu the foreign minister. Their political arguments and strategies are formulated within the framework of international law and agreements, including the United Nations Declaration on the Rights of Indigenous People. They are treated with cautious respect as governments either flounder or respond with faltering steps to less radical calls for recognition and treaties.

Margaret and Paul, as the Gimuy Pilgrims, asked Murrumu what Yidindji would want to see in a ceremony remembering the Frontier Wars. They were suggesting that it take place on Anzac eve. They had some ideas borrowed from the events in Canberra: like the banners carried by Anzac Day marchers since 2016, listing the dates and places of violent conflicts on the Australian colonial frontier, right up to the 1940s; like the handmade paper lanterns, one of the features of the peace vigils held with A Chorus of Women atop Mount Ainslie on Anzac eve, creating an ambience of quiet reverence and beauty.

A working group, including Murrumu, began meeting in a supportive waterfront cafe, planning the ceremony to be held in the adjacent park. They had craft workshops to make the lanterns and Graeme painted them a striking banner, reading 'Lest We Forget – The Frontier Wars', showing a silhouetted figure, arms raised like a bugler, blowing the wind of change.

Yidindji wanted red-tipped feathers, used traditionally in ritual, and put forward names of ancestors to be remembered; these were printed on signs to be dotted around the park. Elders would speak and sing, young men would dance their stories; other Indigenous groups would be invited to witness the ceremony; non-Yidindji artists would respond in the spirit of lament with dance, music, poetry and performance. Margaret would play the viola.

The Pilgrims negotiated with landholders, authorities, the RSL, asking for support or at least non-intervention. They got non-intervention. The 'double healing' ceremony started at dusk and ended in darkness, with everyone sharing by lantern light a feast of tropical fruits, recalling the bounty of the forest. After the feast they marched with their peace lanterns to the *Rainforest Shields*, a work by Yidindji artist Paul Bong, installed in central Cairns.

The official Anzac dawn service took place the next day at the cenotaph on the esplanade. It shared features of those around the country – the solemnity as everyone gathered in the darkness but with the Far North adding its own character as a soft rain fell, the light gradually coming, shimmering on the harbour, the birds waking in the palm trees. School children were there en masse, war widows, the defence forces stationed in the region, politicians, a wide range of government agencies, police, ambulance, the judiciary, Border Force, universities, and civil society clubs and associations. Led by the RSL, it started with an acknowledgment of traditional owners; during the proceedings an Aboriginal man performed on the didjeridu; among many others, a representative of Indigenous veterans laid a wreath. There was no mention of the Frontier Wars.

Margaret made a video about how to organise a Frontier Wars ceremony, urging other people to organise their own next 24 April. In 2017 more people in Gimuy got involved, including theatre-makers. 'It's fantastic!' one of them told Margaret, having witnessed the 2016 event. 'But one little thing, you can't see and you can't hear!' They brought in lighting, amplifiers and stagecraft, to great effect, but the key remains 'relational organising', says Margaret. First and foremost, that means listening to, then following the directions of the Yidindji elders. They supported a new element for the 2019 ceremony, the laying of wreaths to remember Yidindji ancestors tortured and killed on the frontier. The wreaths were laid at the *Rainforest Shields*.

In 2019 the ceremony also emphasised 'truth-telling', as a process 'requested by Aboriginal people across the sovereign nations' following the Uluru Statement from the Heart. In Gimuy they remembered particularly the first post-contact battle in the central Walubara Yidindji area, fought in 1868 at Smiths Creek on the inlet, not far from where they were gathered.

Alice Springs is well behind in anything comparable. Extensive local war memorialising sprang up during the First World War centenary years, on and around Anzac Hill and at the cemetery, yet there is no public memorial anywhere in town to frontier contact, violent or otherwise. By contrast, a number of men remembered in local street names are known to have carried out punitive and reprisal killings of Aboriginal people. In 2010 a giant statue of the explorer John MacDouall Stuart was erected to celebrate the 150th anniversary of his first expedition into central Australia. Its depiction of him as a solitary hero with a gun caused controversy that has not completely gone away, but neither has it led to any action of much consequence. In 2019 there was a multi-faith Frontier Wars remembrance ceremony held on Anzac eve but no Aboriginal people took part. The next day, the traditional Anzac observances were as strongly attended as ever, including by local Aboriginal veterans and horse riders from the Aboriginal community of Ntaria (Hermannsburg) in honour of Australia's Light Horsemen, some of whom were Indigenous.

When the Pilgrims were in Alice Springs for the convergence in 2016, they held a far more elaborate ceremony on Anzac Hill, a 'Lamentation in Landscape' as they called it, for the dead of all wars, including the Frontier Wars. The hill looked beautiful in the early evening light. Graeme Dunstan had arranged lanterns either side of the obelisk, with long narrow banners fluttering above, showing the dove of peace. A man stood between them holding a large Aboriginal flag; a small fire was burning in a metal pan in front of him. More banners flew from high poles around the edge of the hilltop, showing the red-tailed black cockatoo rising with the sun, this symbol of the convergence all but eclipsing Pine Gap's radomes. Margaret and Franz played the lament they had composed for the trespass action, a more polished performance this time. Others sang; someone read a poem.

Graeme spoke about what they were trying to do. How can the peace movement gain recognition for a more inclusive remembrance? he asked. The challenge is to somehow access the 'extraordinary cultural phenomenon' around the commemoration of Anzac Day. In Canberra that year, 140,000 people had risen before 4 am to be at the Australian War Memorial's dawn service, a huge crowd. What is driving this resurgence, Graeme asked, when acknowledgment of the day had almost died during the '70s? One reason, he suggested, is the widespread interest in genealogy; another, providing the Pilgrims more to work with, is people's desire to experience the sacred, 'standing in a big crowd, all feeling the same stuff'. That 'search for the sacred' is being 'hijacked by the RSL, turned into a glorification of noble sacrifice', he said.

He didn't expand on this point, he wasn't needing to convince anyone present – mostly peace activists and sympathisers – but the two speeches at the official Cairns ceremony that year provided an example of this elevation of sacrifice. Both were about the Battle of Fromelles in France one hundred years ago. It was evoked as the 'most tragic event of Australian military history', causing the 'unnecessary loss of so many sons'; it was 'pointless, senseless, futile'. The soldiers knew they faced 'almost certain death', yet still they went, as ordered, over the top of the trenches. For a while the speakers, from the RSL and the Navy, sounded as though they were on the same page as the peace movement. Where they part ways is in the honouring of this 'sacrifice' as an ideal of human achievement, while the peace movement stays with the tragic futility and wants it to stop.

How do the stories of the same events lead to such different understandings? The answer must lie in the way we individually make sense of our experiences and emotions, as reflected in Margaret's intuition about drawing on the personal experience of grief in the

Pilgrims' peacework. I think of my father and the long shadow that the Second World War cast on his life. His brother Brian, the second-born into a family full of sons, seven of them, and two daughters, joined the Air Force. He was killed flying over Denmark on his first mission, in April 1943. He was twenty-three years old and much loved in the family. Coming on top of a younger son's death due to illness in 1940, the loss of Brian broke his mother's heart and health. She didn't try to stop my father from following him into the Air Force. He was still overseas, in training, when she died suddenly in November 1944. She was fifty-two years old.

I visited the Australian War Memorial for the first time in 2018. Like many, I searched the Honour Roll for the name of the uncle I never knew. He hadn't had time to make much of a military contribution – by dropping bombs, for that is what he was trained to do – so it's hard not to think of his death as futile, especially when it cost his family so much. Only his brother Paul, the youngest in the family, is still alive. I asked him how he thinks of Brian's death. 'Heroic' was one of the words he chose, 'not least because he knew the chances of getting through were not good – I heard him say so to my mother before he left for the ship'. The other was 'wasted'. Like a number of our family members, Paul and his wife, Louise, had visited the cemetery in Esbjerg, Denmark, where Brian is buried: 'We could not fail to notice how young were those who lay there,' he wrote, 'including Germans and Austrians; death did not discriminate. All those young folks, and all the years of grief their loved ones bore.'

On the Honour Roll, Brian's death receded dizzyingly into the mass of deaths. I retreated to go looking for the stories of mothers of dead sons and found some in the museum's displays.

The peace movement must create its own secular sacred rituals, Graeme Dunstan told the crowd on Anzac Hill, before he introduced Vince Forrester, a Luritja and Arrernte man, an 'activist in his own community here' and a Greens candidate in the Northern Territory election just passed, who would speak about the Frontier Wars in Central Australia. The sense of the sacred retreated somewhat. Vince spoke with anger, understandably. 'The conqueror had the sword,' he said, 'but he also had the pen, writing us out of history in our struggle.' He urged his listeners to 'go looking for the shield that has the bullet holes in it': through them 'you will see the continuing result of the Frontier Wars'. He spoke of occupation, 'whether it be excluding people from being counted as human beings, taking over our water for cattle. And of course people fought back.' In this history, people in central Australia, 'my ancestors', he said, were the last contact: 'White fella in the history books calls this one "cattle killing times", Anangu say "rifle time". That rifle time went from the 1880s right through to the 1940s.' And it's not over, he said, putting the poverty in bush communities, the destruction of language and culture and the 2007 federal Intervention on that same continuum of destruction.

~

The hearing of charges for the Toowong Cemetery action was set down for 19 July. The good news was that it would be in the magistrates court, and not a jury trial. Supporters arrived from various parts of the country and some came from New Zealand, gathering at Dorothy Day House, 'about twenty-five people from different places and generations, breaking bread and sharing life together', as Andy later recalled. With such solidarity behind them, he started to feel excited about defending the case, which they would do themselves with a bit of advice from a lawyer friend.

When the day arrived, the Christians, as they referred to themselves in this action, and their supporters, a crowd of about forty, walked in procession from Emma Miller Place, filing into the courtroom in silence, holding aloft signs and banners. 'Put the sword on trial', 'Live by the sword, die by the sword – Jesus', 'Mourn the dead, heal the wounded, end the wars'. The public gallery was soon overflowing and people were sitting on the floor.

Chief Magistrate Ray Rinaudo moved the hearing to a bigger courtroom. He rejected submissions by Franz and Andy that they had no case to answer. Checking the *Queensland Criminal Code*, he said the evidence showed they were willing participants at the event. The four then pleaded guilty to the removal of the sword but not to the damage.

Some of their case, the reasons behind their action, was covered in the prosecution's evidence. Jim, for instance, had told the police he had not sought permission from civic authorities to remove the sword: 'Hopefully, there was a higher power directing me.' In a radio interview played for the court, it had been put to him that many veterans feel that the Cross of Sacrifice honours their service, which can involve a 'blood sacrifice'. Jim's answer was unwavering: 'The sacrifice of Jesus was laying down his life for our sins. He didn't take up the sword, he told disciples not to take up the sword. That message is right throughout his teaching, love your enemies, do good to those who hate you, turn the other cheek, etcetera.' This is why having the sword on the cross, in the place of the body of Jesus, was such a shocking betrayal, he explained. On air, his voice was calm and restrained, his sincerity seemed absolute: he was prepared to go to gaol if it came to that; he had also been prepared to pay for repairs to the cross but only if the sword were not replaced. It already had been.

Sincere belief in the rightness of what they were doing, restoring the cross to its proper sword-less state, was the essence of their defence. To this end they called their first witness, Dominican priest Peter Murnane, who pushed back at vigorous questioning by the prosecution to outline 'the symbol of the cross in Christian theology and how it was incompatible with that of the sword'.

A second day had to be scheduled, with the weekend intervening. Some supporters returned home but there was still a good crowd in court. Now it was for each of the defendants to take to the stand and speak for themselves. They planned to make the most of it. Franz was going to sing 'The Vine and the Fig Tree', the same anti-war hymn he had sung at the cemetery, but somehow that didn't happen. Andy was going to read from the Bible: 'When I was inevitably interrupted I would say "that's just it – the Bible's used as a prop to give human institutions some kind of divine legitimacy, but you're not allowed to mention what it actually says." In the end though, under pressure on the stand, I chickened out and let that golden opportunity slip away forever.'

Their conviction for the offence came as no surprise. 'While we aspire to religious freedom, we must condemn actions where that is taken to the lengths of destroying public property,' Chief Magistrate Rinaudo said. 'You simply cannot do this with impunity and think you can hide behind your beliefs. It doesn't work like that.' Not that any hiding had been involved. Jim and Tim had always been completely prepared to own this action.

The sentences, delivered the same day, were reasonably lenient: Jim got a three-month wholly suspended jail sentence, plus 100 hours of community service and was ordered to pay restitution for half of the $17,812 damages bill. Tim was told to pay the other half and got the same community service order. Franz was fined $1000 but no conviction was recorded. Andy was convicted and fined $1500. Jim's community service was later rescinded as it exceeded the magistrate's powers to impose two penalties.

Outside the court, Jim defended their action in the tradition of civil disobedience but also apologised to veterans who were upset by what they had done: 'It was never our intent and we meant no disrespect to veterans.' Then they went across the road to Emma Miller Place for a relieved picnic – no pub for the Catholic Workers.

'We went to bed that night free,' Andy wrote in his blog many months later, 'not because of the benevolence of the state (something we were admittedly grateful for), but because we had resolved to do what we believed was right and would continue to do so through good and bad circumstances.'

He still felt conflicted, but valued the way the whole extraordinary experience had reminded him of what the cross actually means: 'It was the method of execution a court somewhat similar to ours passed down on another criminal who challenged the religion of the day … Another quote from Jesus to his disciples: "If the world hates you, keep in mind that it hated me first."'

~

The Pilgrims' jury trial for the Pine Gap trespass was quite close now. Russell Goldflam, as their legal advisor, had been trying to get them to focus on preparation; he'd been trying to get the Commonwealth Director of Public Prosecutions to do the same. Not even the indictments, listing the charges, were to hand.

'The Pilgrims all wanted to plead not guilty, but they also all wanted not to go to gaol,' he recalls. 'They were worried that gaol was more likely if they pleaded not guilty. They were also uncertain about whether they would admit certain facts or put the Crown to proof on every single element, which would blow out the trial for weeks. So I was asking the Crown to give notice of what facts they were wanting them to admit. That took ages.'

Finally they all decided to plead not guilty while admitting the basic facts. Things began to fall into place. Russell would continue to give them legal assistance for the pre-trial hearings but, when they didn't get legal aid approved, they got ready to represent themselves in the trials.

A RECKLESS ACT OF PRAYER

'There are no really analogous cases. It's a unique Act as far as I know, with a maximum penalty of seven years for what is really a trespass offence.'

That didn't stop Crown Prosecutor Michael McHugh calling for a custodial sentence for Paul Christie, he who was arrested inside Pine Gap with a rattle and flowers in his hands. He had looked rather vulnerable as he entered the courthouse alone, turning back to say a few words to his friends outside, who had been singing peace songs and praying. Dressed in black, long pants, long-sleeved shirt, his hair combed slick, he was holding aloft a flag on a pole, showing the planet, its oceans and continents and atmosphere, in a field of deep blue, making a connection between his small action and the big picture – the world.

Paul's trial came first. The Pilgrims had applied to have a joint trial but when the Crown opposed the application, the court refused it. Now McHugh's submissions on sentence foreshadowed what they all would face. Paul, though, had mounted no real legal defence to the charges, whereas the five would – they would try anyway.

13 November 2017

Paul asked for Graeme Dunstan to be his 'McKenzie friend', sitting alongside him to give moral support but not allowed to otherwise intervene. Justice John Reeves would assist to a degree with procedures but his help would be limited. He started by explaining Paul's rights in relation to empanelling a jury – they included having six opportunities to challenge, while the Crown had twelve.

Jury empanelling is long-winded but it can offer a glimpse into the community dynamics around particular cases. This one was no exception. After Reeves had heard three whispered requests from potential jurors seeking to be excused, he spoke to the 'muster': If they worked at Pine Gap, or their husband or wife did, or they knew witnesses who did, he would not regard that as 'private'. It was important as far as possible for everyone to hear why some people were being excused.

Two calls later a woman said she was a former employee at the base. Reeves thought it best she didn't serve. She was followed by a man who was a current employee of the base. Excused.

And so it went. Fourteen people in all sought to be excused. Of those whose reasons we heard, four were current or recent former employees at the base. A fifth had left his job there ten years ago. Would it affect his capacity to listen to the evidence impartially? He didn't believe so. Not excused. A sixth had worked at a similar base elsewhere in Australia; he'd left the job about nine years ago but he couldn't say he was unbiased. It was so long ago that Reeves would not excuse him. Two people were friends of one or more of the witnesses involved in security roles at the base. They were excused.

Obviously all these people, randomly selected, were Australian citizens. It was a snapshot of the reach of the base into the local community – even without the Americans.

Another snapshot was Justice Reeves's own past association with Pine Gap. As a young solicitor he had lived in town. He'd served as a town councillor and got involved with a community group raising local awareness about Pine Gap and its activities. The Pilgrims had no problem with this, as their advisor Russell Goldflam told the court in a pre-trial hearing on 16 August. It was the first the Crown had heard of the issue and at that stage nothing more was made of it.

On 9 October, a little more than a month out from the start of Paul's trial, that changed. The Crown would be presenting materials on the issue, advising of the 'potential for apprehended bias'. The materials included two articles from the *Alice Springs News*. One, from March 2002, quoted Philip Nitschke, who would later become a doctor and a household name as an advocate for voluntary euthanasia. In 1976–77 he had been working as a ranger at Simpsons Gap and as caretaker for the Temple Bar caravan park in Ilparpa Valley, just south of town. From the top of the hill there you could get a really good view of the base: 'I had an almost continuous stream of people who wanted to have a look, so I was more or less doing tours up to the same spot on top of the hill.' He knew of Des Ball's work on the base and he'd ring him up with news about what was being built there. 'That was when we formed the group Concerned Citizens of Alice Springs ... John Reeves was incredibly important at that stage.'

The second article, from November 1998, was about Aboriginal land rights; the reference to Reeves was because he had authored a review of the Land Rights Act. By way of background, the article said he had been, among other things, 'a vociferous opponent of the US controlled Pine Gap intelligence gathering base here'.

A third item, an extract from Nitschke's memoir, had Reeves centrally involved in the beginnings of the ginger group: 'With lawyer John Reeves (now a judge of the Federal Court), I founded the group Concerned Citizens of Alice Springs and began publishing a regular newsletter critical of the American presence there. With others I occasionally camped outside Pine Gap and probed the perimeter, and we constantly voiced our concern about the secrecy of the installation.'

The Crown was not making an application for Reeves to disqualify himself 'at this stage'; they had provided the materials 'for the assistance of the court'. Russell, for the accused, expressed concern about the Crown's equivocation: it left open the prospect that at this late date the trials might be aborted.

Because Reeves is an 'Additional Judge' of the Northern Territory Supreme Court – since 2007 sitting mostly in the Federal Court in Brisbane – the pre-trial conferences were being heard by Justice Peter Barr. He described the conduct of the Crown on this issue as 'cavalier': tendering the material without making an application was of no assistance to the court. Ten days later, counsel for the Crown, Liesl Chapman SC, advised that they would not seek the disqualification of Reeves. That seemed to settle the matter until, on 2 November, the Crown advised that Chapman would no longer be available for the trial.

The next day, 3 November, ten days out from the start date, Reeves himself conducted a pre-trial conference and made a statement about the issue. He confirmed what had been tendered and went further: he had been involved in a group in the late '70s which had concerns about Pine Gap; he had read what was available on the subject at the time, including Des Ball's book; he had attended a Labor Party National Conference and participated in debate about Pine Gap; he had never been in any protest against Pine Gap, or been near the fence, or camped outside the gates; but he had been with groups who had climbed a hill near Temple Bar and taken photographs of the base. In response, the Crown wanted more time, keeping alive its equivocation. Reeves coolly requested that if an application that he disqualify himself was to be made, it be done sooner rather than later.

The Pilgrims noted the irony in Reeves's admission of having taken photos of the base, prohibited under the *Defence (Special Undertakings) Act*. Maximum penalty: imprisonment for seven years.

The calendar ran down to the start of Paul's trial; no news from the Crown on Reeves's 'apprehended bias'. The Pilgrims did not even know who would replace Chapman as Crown counsel. If an application for disqualification came up, which Russell still considered to be a real possibility, he would appear on their behalf for the argument.

In the end nothing further was made of the matter. A jury was empanelled, Reeves was on the bench, and Michael McHugh SC was appearing for the Crown. He's a tall man with a rather patrician face and bearing, but his manner was relaxed, even friendly. The self-represented defendants, though, would not get away with much and when it came to sentencing submissions he was ready to play hardball (as his client no doubt required). He was assisted by a 'junior', Caroline Dobraszczyk, an unsmiling woman of mature age and significant legal experience. An instructor, a small bespectacled youngish man named Myles Gillard, handled their voluminous lever-arch files. Sitting behind this team and frequently leaning forward to speak to them were two representatives of the Commonwealth, at least one of whom was a lawyer.

No wonder Paul had asked for a McKenzie friend. Still, he had plenty of supporters in the public gallery. Soon they would include his father, David, arriving from Sydney mid-morning. They're close. Paul talks of both parents having laid the foundations for the way he has lived his life. His dad is an engineer who chose to work for the state rail authority 'because rail is the best form of mass transport'; he rose to the position of chief engineer. His mother, who died in 2001, was a doctor who worked in family planning. They were both from Baptist families, with relatives who were missionaries on both sides, which Paul mentions as a measure of the strength of their religious commitment. He has moved a long way from this heritage. 'God save us from religion!' he will say as a way of explaining he is 'not anti-God, just anti-religion'. As for his parents' modelling of good citizens' lives, he respects it but sees them as duped by 'the imperial narrative'. The bubble burst for him while he was still at school and discovered, through his own reading, differing accounts of the war in the Pacific: the Japanese were preparing to surrender before the atomic bombs were dropped on Hiroshima and Nagasaki; he'd only been told that the Americans, Australia's allies, were 'the good guys' – yet they had done this. 'What other lies was I being told?' His search for the answers to this question has shaped his life ever since, leading him to the 'beautiful social experiments' of Rainbow Gatherings, communal living and activism.

Jim and Franz were prevented from entering the court because of their bare feet. Jim was not going to back down on this and Franz stood in solidarity with him. They wondered whether court staff would try to stop them coming to their own trial – that didn't happen, of course. When Andy was refused entry, being 'not as staunch a bare-footer', he cycled back to their accommodation and got a pair of thongs, which he brought with him each day after that. Tim also put on shoes so he could go inside for Paul.

Some of the supporters were there for more than solidarity. They wanted to make the most of the exposure the trial would give to their cause. There was a lot of industrious typing for a daily blog as well as tweeting, and actions were being organised for outside the court, including daily praying and singing before the hearings started. Part-advocate, part-scholar Felicity Ruby was in close contact with them. In the lead-up to the trial she had published a piece about the prosecution in *Arena* magazine. A former researcher and advisor to the UN and later to Greens senator Scott Ludlam, she was now doing her PhD on how the Five Eyes alliance operates in Australia as well as on the resistance to surveillance technologies by political movements, the Pilgrims being a case in point.

There were other journalists apart from me: a senior reporter for the ABC, a freelancer who had a piece commissioned by *The New York Times* (the activists were very excited about that), another who had also followed the Christians Against All Terrorism trial, and occasionally a reporter from News Corp's local paper.

Eight women and four men made up the jury Paul faced; a woman was the reserve. They included a midwife, a shop assistant, more than one admin officer, a park ranger, a disability pensioner. None appeared to be under thirty.

It was the first trial I had attended in the new Supreme Court building. It rises well above any other in town, a glazed oval drum of five storeys, the first (and to date, only) local structure built to that height after the old three-storey limit was lifted. At its inauguration just a few months earlier, the chief justice had sought to counteract the unfortunate message it conveys: 'By its size and singularity of design this building might suggest to some that this court seeks to place itself above the community. Any such impression would be entirely mistaken.' Too late by then to correct it.

My more pressing problem now was the acoustics in the courtroom. Reeves has a soft voice and it was getting lost, especially against the rattle of the brand-new air-conditioning, which court staff told me they had been unable to do anything about. Another frustration was the jury box arrangement. Observations of the jury are a feature of court reporting. Are they attentive, riveted, weeping, or eyes glazing over, falling asleep? In this courtroom the jury box is angled away from the public gallery, so the jurors can mostly be seen only side-on, and anyone short in the raised second tier disappears from view behind its solid barrier. I never thought I would miss the hard benches and cramped conditions of the old building, the often visceral proximity of all parties to the trial.

Reeves, from his remote high bench, warned the jury against gathering information outside of the court. When they left the building, they should avoid the area where the Pilgrims had set up office – under a palm tree on the edge of the Uniting Church lawns – and where Graeme Dunstan's Peacebus was parked, draped with the colourful Close Pine Gap banner. They should not visit the scene of the alleged offences (which would have been both difficult and risky, to say the least), not read the media coverage, not use Google Earth. 'Decide the matter with your head, not your heart,' keeping an open mind and not drawing any conclusion until they had heard all the evidence, he told them. The accused was assumed to be innocent of the charges; it was for the Crown to prove them, beyond reasonable doubt.

McHugh outlined what the Crown would do to present its case. The indictment was one count of being in the prohibited area, and the Crown would say the accused was 'reckless' to the fact that it was a prohibited area. Paul would latch on to that word 'reckless' in its everyday meaning, although McHugh was using it in a narrower legal sense. McHugh acknowledged some admitted facts, which would allow a shorter trial, but he spoke too soon. The trial of this matter, 'straightforward at its core', would end up taking three days.

Paul asked to make an opening statement. McHugh was concerned that it would place in front of the jury the possible legal defences, of necessity and defence of another.

Did he intend to speak to those defences? Reeves asked Paul.

Yes, he said, and about his character and the strong beliefs underlying his action.

What, in short, was the necessity he was facing? asked Reeves.

'The escalating emergency perpetrated by the military base,' said Paul, citing the deaths that had occurred only days before his action (the strike on Shadal Bazar), which he said breached the *War Crimes Act 1945.*

And how had he acted in defence of another? asked Reeves.

Operations at Pine Gap had contributed targeting information for those fifteen people who had been killed in Afghanistan, said Paul. He had 'grave fears for others' which is why he had entered Pine Gap.

McHugh was concerned about Paul raising 'matters which will not be justiciable in this court'.

Reeves questioned this. McHugh cited Justice Thomas in her ruling against the defences argued for by the Christians Against All Terrorism, but Reeves countered with a reference to the introduction to chapter 8 of the *Commonwealth Criminal Code.*

I looked it up later. Chapter 8 deals with offences like genocide, crimes against humanity, war crimes. The introductory paragraphs are about Australia's right to exercise its jurisdiction with respect to these kinds of offences that are also crimes within the jurisdiction of the International Criminal Court. Reeves enjoyed this kind of to and fro on matters of law, not that it would ultimately make any difference to how this case was adjudicated. Paul was not about to mount serious legal argument about the implications for operations at Pine Gap under the *War Crimes Act.* Yet McHugh was assiduous in trying to head off anything in that direction. Keep it simple for the jury was his guiding principle, so he would have been pleased that Reeves did not let Paul make an opening statement.

The first Crown witness was Matthew Gadsby, the Protective Service officer who had spotted Paul inside the base on CCTV. A heavy-set middle-aged man, he appeared in his dark blue, shiny-buttoned service uniform. He spoke in short fast bursts, machine-like. He was tense, I realised later when I passed him on the stairs and he asked me in his ordinary voice what time court would resume.

McHugh took him through his evidence, in exhaustive detail. A map was produced and he was asked to mark where Laura Creek begins and ends in its traverse of the base from west to east; where he saw 'the person of interest'; where the firebreak road runs along the fence-line and so on.

Is there a fence encircling the facility? McHugh asked.

Actually two fences encircle it, Gadsby replied.

McHugh asked him to describe them, but quickly withdrew the question. The reason why became clear a little later.

After the arrest Gadsby had sent an officer to photograph evidence of the incursion. He'd also requested the download of the CCTV footage which was shown to the court.

Everyday citizens don't often get to see inside Pine Gap. I made that a focus in my report for the *Alice Springs News* at the end of the day:

> On grainy CCTV vision we saw a sandy creek bed in early morning light, some of it perhaps thick with grass, shaded by tall gums – the kind of place you would be happy to make camp, except that you would likely be arrested.
>
> A small figure, noticeable mainly for his red clothing and gear, entered screen left, walked along the creek for a bit, then turned around and walked slowly back in the other direction.
>
> The time counter ticked over from around 6.26 am on 3 October 2016 to about 6.31, when an AFP vehicle arrived, and perhaps another, along with a number of uniformed personnel.
>
> It was hard to see, through the trees and scrub, the arrest of the 'person of interest' that followed.

Paul started his cross-examination by thanking Gadsby for his service. He was at pains to be courteous in court, almost to the point of deference, but he was also capable of quiet insistence. He asked Gadsby if he, Paul, had posed any danger or threat or had objected to his arrest.

McHugh immediately objected on relevance.

Paul said the answer could go to his intentions, whether they were mischievous or sinister or reckless.

Mr Gadsby's opinion of that was irrelevant, said McHugh, and the question was also irrelevant to the fact in issue, which was unlawful entry in a prohibited area.

That is probably correct, said Reeves, but he would allow Mr Christie some leeway. Did he pose a threat? he asked Gadsby.

Yes, said Gadsby, he was in a prohibited area.

Paul asked whether a red hoodie was the sort of clothing someone with mischievous or sinister intent would wear.

Reeves again told Gadsby to answer.

Yes, he said, you were in a prohibited area where you were not meant to be.

Paul went to the matter of the fences around that area, the outer fence which is a typical cattle fence, and the inner fence. Could the latter be easily penetrated?

Objection, this again went beyond what was at issue.

Paul said it would speak to the nature of his action, its lack of recklessness.

Reeves allowed it.

McHugh was on his feet: he wanted to raise a legal matter. The jury was sent out.

He was the one who had brought up the fence, Reeves told him.

McHugh acknowledged that, but he had not described it. This question about penetrability went to matters of security and was against the public interest, certainly in open court.

So, this was the reason he had withdrawn his own question. Really? Everybody knows there is an inner 'man-proof' fence around the operations compound. Its breach – by a man and a woman, Jim and Adele – during the 2005 trespass is famous. In their trial, open court had even learned detail about the heavier gauge materials used in its construction after Jim had questioned the seemingly excessive cost for its repair.

Reeves asked why Paul wanted to know about the penetrability of the inner fence.

Again, it would speak to the intent of what he was doing at the time (walking back and forth with his rattle and flowers, and no tools in his possession other than a small pocketknife): 'I had no intention or capacity to cause a breach.'

That statement could be put to the jury in his closing address, said McHugh. It did not require a question about penetrability.

Reeves disallowed the question and the jury was brought back in.

Paul asked Gadsby if he had offered resistance at his arrest.

I can't answer that, said Gadsby. (He had not been present.)

Did you hear that I offered resistance?
No answer to that either.

The arresting officer, Lloyd Adams, strode to the witness box, wearing a suit, not his uniform. He answered questions as impersonally and briefly as possible, never looking directly at his questioner. He was shown the map marked by Gadsby. Did the mark in the Laura Creek bed mean anything?

Yes, it was the area where 'the person of interest' was seen on CCTV and where they 'deployed to'.

And so the evidence went, increment by tiny increment, until he and his colleague 'gained subject control of that person'. He 'cautioned that person', placing him in flexi-cuffs 'for officer safety' until he was handed over to the Northern Territory police.

In cross-examination, Paul referred to phrase 'gained subject control'. Did I offer resistance? he asked.

No, said Adams.

Federal Agent Peter Davey, who conducted the interviews with both Margaret and Paul after their arrests, took the stand. He's tall, trim, with a dour expression, at least in court. He talked in a bloodless way, his eyes fixed at all times on a point in the central space of the courtroom. McHugh's first questions went to his role during Operation Fryatt. Davey had been on 'numerous occasions' to both protest camps during the convergence. He had been in plain clothes, but he had told the activists clearly who he was, and he was armed, with gun, baton, spray, a bumbag 'full of accoutrements'.

He had done an area familiarisation, including an inspection of the base fences. McHugh asked him quite a few questions about them, including the fence at the entrance on the southern side, three metres high, chain mesh, topped with six strands of barbed wire, described by Davey as 'quite formidable'. McHugh stayed well away from any questions about the inner fence.

Davey's evidence went on to cover Paul's arrest, his interview, and the investigations, for which he had produced a meticulous record: a Google Earth image of the area overlaid by a survey map and checked for accuracy by a geospatial services officer. McHugh tendered the document.

When I cover a trial I report on each day's events. It makes more sense of the process to follow it as it unfolds and should make more sense of the verdict and sentence. Who knows, though, if that's how readers take it? Those who make public their responses tend to stay in their camps irrespective of what transpires, but this in itself is a certain barometer of local attitudes.

A regular commenter on local affairs goes by the pseudonym 'Fred the Philistine', a fair reflection of the general tenor of his comments. He surprised me on this story: 'This needs to be brought to attention,' he wrote. 'America is not innocent. They are behind a lot of the war destruction happening in the world today.' Another commenter under the self-explanatory pseudonym 'Close the Gap – Close Alice Springs', kept it pithy: 'Economics Rule', reflecting the expedience behind the widespread local acceptance of the base. A third, who would continue posting on reports of both trials, was John Bell, his real name. He wrote: 'The protester wants notoriety and has got it. The PSOs had a job to do and they did it. The protester is representing himself as is his right. We know the old saying that a man who represents himself has a fool for a client. The Commonwealth is represented by silk as is its right. The only real interest left in this case now is whether the court will award costs. That's the killer – and I speak from experience.' He was wrong on that last point. He had represented himself in a civil case and lost, but costs are never awarded in jury trials.

14 November 2017

Paul's interview with Davey on the day of his arrest was played. The account he gave of his long walk into the base left an impression, on me at least, of his resolute character as well as physical hardiness. His world view was clearly a long way from mainstream, but it was gentle and compassionate. Who knows what the jury made of it, including his claim of having the 'permission, blessing and support of Arrernte elders' to be on their land. Even people respecting the fundamental right of Arrernte people to call this country theirs may have wondered about Paul's sense of permission. Its public face was Chris Peltharre Tomlins who was often in court during the trials and at the actions outside the court. He is not unknown in the public life of Alice Springs and also has a profile in alternative lifestyle and political circles on the east coast. He had put out a call in these circles for people to join the 2016 convergence and was a prominent presence at the Healing Camp. I later asked Paul more about the permission he felt he had. Tomlins was not the only local Aboriginal person he met at the Healing Camp, he said, naming other names, but he was the main connection.

This issue, though, was not the one to bite in the local response to the case as I reported it. Rather it was Paul's description of the base as 'illegal', as carrying out 'murderous activity', as 'one of the biggest war criminal hotspots in Australia', and of himself as 'just one man, facing a machine of death'.

'Dramatic and evocative language from the Peace Pilgrim but not factually incorrect,' wrote a man under the pseudonym 'War machine'; he posts occasionally, under different pseudonyms, mainly on environment and social justice issues. 'The facility at Pine Gap means that Australia IS complicit in war crimes committed in the Middle East against civilian targets. It is absurd that during a time of "peace" there is an active military base here in the centre of Australia. The American war machine may seem invisible but its influence is all encompassing.'

'I agree,' posted 'Come in spinner', who has a similar range of interests, 'he's essentially a good guy who acted with honourable intention and sound ethics.'

John Bell was less sympathetic: 'Does Peace Pilgrim have the right to choose which laws he can obey and disobey? Is there a limit to how far Peace Pilgrim can extend the middle digit of protest to lawful authority on any cause he thinks is ethical and honourable? How far is Peace Pilgrim lawfully allowed to push the limits of our standards of free speech before a fair-minded objective observer would say barley charley, fair suck of the sauce bottle mate?'

They were points that both prosecutor and judge would later take up, albeit in more decorous terms.

~

Before then there was more painstaking proof of the fault elements to get through, including the photographic evidence, complete with GPS coordinates of a distinct set of bootprints all the way to the western boundary fence, and the admissions of fact, which covered in brief everything that we were hearing in laborious detail.

Finally, it was time for Paul's evidence. He still wanted to raise the legal defences, in the most general sense: he wanted to speak about why he felt justified to act as he had and why he didn't see it as reckless.

Reeves tried to get him to a point where he could either outline the evidence underpinning the defences so that Reeves could make a ruling, or go into the witness box and give his evidence under oath. In the end Reeves decided the latter was the safest step in the circumstances. He would then make the ruling and give directions to the jury afterwards.

Paul did not challenge the facts presented by the prosecution, but rather attempted, with a 'bit of nervousness', to offer another perspective on them. He hadn't been able to plead guilty out of respect for his own 'nonviolent peaceful prayerful action', his sense of having done what was right. He believed in the rule of law when it was seeking to defend the rights of all people, but he reserved the right to oppose violence wherever he saw it.

He believed the ownership of this land (including Pine Gap) was in question: sovereignty by First Peoples had never been ceded. He spoke of his work in Cairns with the Yidindji Sovereign Nation, saying that he lived for the day when Australia would conclude a treaty with First Peoples 'so we can move forward'.

He had been called to the Centre by Arrernte people. He read out loud the invitation from Chris Peltharre Tomlins. Addressed to 'all the peoples of the world', it spoke of the 'immense' harm caused by Pine Gap and the intention in 2016 to close it: 'The Australian Government is incapable of doing this, so we the Arrernte people must take responsibility. We need your help ... Come to the land of the Arrernte to heal, to live together, to share culture and build a positive future.' It spoke of a 'spiritual awakening', of 'healing and smoking our spirits'. Paul said he had no way of fighting his personal drive to come in response to this call. Once here, he had met 'many other Arrernte people', they had been 'welcoming and encouraging', giving him 'their blessing'.

He anticipated that the prosecution would show the jury his 'long police record'. Its length, I saw later, was mainly due to traffic offences; there were a couple of cannabis offences from the early '90s; the most substantial fine was $500 for his trespass on Commonwealth land during Talisman Sabre in 2015, with no conviction recorded. Paul 'humbly' acknowledged the 'mistakes and oversights' his record revealed, but its political actions spoke to his character, he said. He had worked hard to do what he thought was right, including the community development and social welfare work he had dedicated himself to, in 'support of others, in support of the meek', never for enriching himself. He had no assets or wealth to speak of in the financial sense but felt 'blessed and supported' in the work he did, which had taken him 'on a journey that hasn't finished'.

He took issue with the word 'reckless' in relation to his charge: Pine Gap's involvement in unlawful, unsanctioned operations, escalating the war on terror, 'that to me is reckless'.

He was sure that 'each of us' must do something. For him it was walking on the land at Pine Gap and praying for Arrernte children and all the others caught up in the 'out of control war on terror and its disastrous assassination program in the Middle East' for which the targeting was contributed to by Pine Gap. This he knew from information in the public domain. He would not be able to bring evidence of that to court, he would be 'shut down' but, he argued, a nation should be transparent when it comes to taking the lives of others and that was not the case with Pine Gap. The Australian Government had never 'verbalised legitimate reasons' for the operations carried out at the base.

He reminded the jury that he had at all times been polite and respectful, before asking finally, 'When is a prayer a reckless act? When is singing for children, singing for peace, a reckless act?'

The thrust of McHugh's cross-examination was to get Paul to admit that he knew he was in a prohibited area without permission, that this was his intention and that he knew it was breaking the law. He didn't go in hard, he didn't really need to, although occasionally Paul managed to side-step him. He put to Paul that he had given evidence (in his interview with Davey) that he knew there was a sign saying 'Stop' at the front entrance.

His interpretation was 'a bit different', said Paul. He saw it as a sign to the base to 'stop their activities'.

Paul agreed that he was aware that some Peace Pilgrims had been arrested a few days earlier, but he wouldn't adopt the term 'decided' or 'intended' in relation to what he had done – he had been 'called'.

He did eventually accept McHugh's distinction between a written permit and the kind of permission he felt he had from Arrernte people to be on their land; he agreed he did not have a written permit.

The purpose of his action was to protest, put McHugh (the implication being that it was not in response to an extraordinary emergency and not to defend self or others – the legal defences Paul had hoped to raise).

It was to pray, said Paul.

Couldn't he have done that at the front gates?

He could do it anywhere, and yes, at the front gates.

He wanted to raise debate about the future of Pine Gap, put McHugh.

That was part of what he was here for in 2016, part of what he was still here for.

He knew the area was a prohibited area, put McHugh.

He understood that Commonwealth legislation described it as such, answered Paul, and he agreed he knew that on 3 October 2016.

He had spoken about obeying some laws and not others, put McHugh.

Many laws need to change, said Paul.

But that could be done through the parliament, right?

History shows, said Paul, that agitation by people in the community gets laws changed.

McHugh put to him his statement that he reserves the right to obey some laws and not others.

Paul hesitated, then said 'as a human, yes'.

And that was what he was doing on 3 October 2016, wasn't it? Praying, walking on land that was governed by a law he was breaking.

Yes, accepted Paul.

He left the witness box. He had no other witnesses to call, no other evidence to tender.

When Reeves returned from afternoon tea, with the jury out of the room, he heard from the Crown on the legal defences. They had not been raised, contended McHugh. The accused had not tied his trespass to any extraordinary emergency, nor to defence of another, except for a generalised prayer for children. His Honour could simply give directions to the jury to look at only the evidence and ignore submissions.

Reeves was not ready to so quickly dispense with the matter. In dealing with a self-represented defendant, it was clear that he wanted to be demonstrably fair. But the further to and fro with McHugh simply covered the same ground. Reeves turned to Paul, who now said that he was not pursuing those defences, that he hadn't done the work to do so.

The jury returned and McHugh started his closing address. He didn't propose to rehearse the 'reasonably fresh' evidence, although soon he would do just that, but first he had some comments to make about the rule of law. Lawful protest is a hallmark of a great democracy, he said, but there are always limits to freedom of speech: one cannot yell 'fire' in a crowded theatre. And one cannot choose which laws to obey.

On the facts, the accused had said he hadn't seen the fences and signs during his long walk into the base, but clearly he must have, said McHugh: he knew or was reckless with regard to that knowledge. Having permission from Arrernte people to go onto the land was not the same as having a permit under the Act, and the accused knew he didn't have one or was reckless in that regard. He had referred to mercy, but that didn't go to the question of liability, McHugh told the jury: their role was to look at the evidence of liability.

Paul closed by answering McHugh on the rule of law. There are times when laws get broken in service of higher laws, he said. Many laws don't change until people start breaking them. He had 'concern and respect for this place' – the court – but he had answered a higher calling, he had no choice. McHugh had said you can't yell fire, but 'there is a fire'.

15 November 2017

It took the jury little more than half an hour to reach their unanimous verdict of 'guilty'. With 'no defences raised on the evidence', as Reeves had put it, there was no surprise in it. Paul was calm, he was glad that it was over. He asked for sentencing to be expedited and the Crown had no objection.

With the verdict in, there was a request from media to access exhibits – including the CCTV footage of Paul inside the base. McHugh would seek instructions. It was the first matter dealt with when the court reconvened: the Crown opposed access to any of the exhibits requested, and particularly the CCTV. They were concerned that it would be an offence under the Act to use the footage of the prohibited area, notwithstanding that it had been shown in court; they also had security concerns about the offending action being shown for any purpose other than forensic.

The Crown's sensitivity about footage inside the base would assume far greater proportions in the next trial.

McHugh began his sentencing submissions. Now there was a surprise in store, in fact more than one. First, he told Reeves that the Crown had consented, early in the process, to have the matter dealt with in the lower court; the accused had not consented and that was an adverse factor to take into account now. Having a jury trial in the Supreme Court was part of his protest; it had imposed a burden on the state and the community, but that cut both ways: a trial in a higher court also meant exposure to harsher sentencing regimes.

All that was news to Paul. He had had no idea that the lower court option had been available to him; he would have to check with Russell.

Reeves asked McHugh what guidance there might be on penalties for similar offending in a similar court.

On this McHugh couldn't really help because of the uniqueness of the legislation. This was when he acknowledged that it was 'really a trespass offence' but, he said, the legislature clearly took the view that the offending in a prohibited area was serious. In this case, it had been deliberate and calculated. As it was contested, there could be no discount on plea, although there was utility in Paul's cooperation with the investigation and his admissions by the time he got to trial.

Then came the bombshell: specific deterrence, he said, put the offence into a category for which there was no appropriate sentence other than custodial. Looming even larger was general deterrence: people must know that the *Defence (Special Undertakings) Act* carries a harsh maximum penalty and if you breach it, then you can expect to be sentenced accordingly. He cited *The Queen v Saleh*: 'The real bite of general deterrence takes hold only when a custodial sentence is imposed.' This is particularly the case with white-collar crime, he said, evidently seeing it as the closest analogy to offending by nonviolent protesters.

McHugh referred to Paul's written submissions in which he had cited comments on civil disobedience by the British Law Lord Leonard Hoffmann. The context for the comments was interesting (not that McHugh went into this). They were made during appeals by several groups of activists, twenty individuals in all, women and men, who had attempted to disrupt war preparations (some of them partially succeeding) in the weeks leading up to the invasion of Iraq in 2003. All of them had done, or intended to do, more than trespass. The Fairford Five, for instance, in three separate actions, had broken into an RAF airfield where American B-52s were stationed; two of them did considerable damage, disabling fuel tankers and trailers used for carrying bombs; another two were intending to clog the B-52 engines with bolts and screws; one, who acted alone, was charged with attempted arson.

When Lord Hoffman was writing his opinion, the Fairford Five had not yet been tried; their appeal was against rulings by judges in lower courts on the defence that they had used reasonable force seeking to prevent the crime of aggression, in other words an illegal war. The legal question the Lords were considering was whether the crime of aggression in international law could be considered a crime in the domestic law of the United Kingdom. Their conclusion was that it was not, but the activists were allowed to contend – as had already been acknowledged in lower courts – that they were seeking to prevent the commission of war crimes. The difference turned on, to put it very simply, what had been defined by the parliament as criminal. War crimes had been dealt with in the UK's *International Criminal Court Act 2001*. However, the crime of aggression had not been the subject of UK legislation, at least in part because there was no international agreement on a legal definition for it. The Lords were also wary of the courts reviewing matters of foreign affairs and deployment of the armed services, seen to be the prerogative power of government, not the judiciary.

When the Fairford Five were tried, juries dealt variously with them. Their first round of trials all ended in hung juries. At their second trial the two who did actual damage were convicted. One of the issues was that the timing of their action, a week out from the invasion, had made it harder to argue that they were acting to prevent an imminent threat to 'life, limb and property'. The pair who intended damage were acquitted at their second trial; and the man who acted solo was also cleared when his second trial ended again in a hung jury. In the other cases the Lords were considering, the appellants had already been convicted and sentenced: they had been discharged, some with conditions or fined, and some compensation orders were made. It was in this context that Lord Hoffman made his comments on civil disobedience:

> My Lords, civil disobedience on conscientious grounds has a long and honourable history in this country. People who break the law to affirm their belief in the injustice of a law or government action are sometimes vindicated by history. The suffragettes are an example which comes immediately to mind. It is the mark of a civilised community that it can accommodate protests and demonstrations of this kind. But there are conventions which are generally accepted by the law-breakers on one side and the law-enforcers on the other. The protesters behave with a sense of proportion and do not cause

> excessive damage or inconvenience. And they vouch the sincerity of their beliefs by accepting the penalties imposed by the law. The police and prosecutors, on the other hand, behave with restraint and the magistrates impose sentences which take the conscientious motives of the protesters into account. The conditional discharges ordered by the magistrates in the cases which came before them exemplifies their sensitivity to these conventions.

It was an eloquent argument for leniency in sentencing, but McHugh felt it was 'not so apposite' in this case which turned on its own set of facts and had been tried under unique legislation governing trespass on a prohibited area for defence purposes. Giving any credence to a breach of the law would also not reflect the jury decision, which had been 'if anything, swift'.

On comparable sentences, McHugh couldn't get away from Justice Thomas's relative lenience towards the Christians Against All Terrorism: the maximum fine was $1250, when the incursion had been more serious and there had also been some property damage. However, the Crown had held that those sentences were 'manifestly inadequate', said McHugh, even though the appeal on them had been abandoned. He maintained that sending this offender to prison was the only appropriate option.

What reasons would I give? asked Reeves.

The legislation provides for a maximum of seven years for this type of offending, said McHugh. The offender's was not the 'worst case' but it was somewhere in the mid-range.

The mid-range? repeated Reeves, with a hint of incredulity. He put to McHugh that the maximum of seven years, on the face of it, appeared to be directed at conduct that compromised security. Paul Christie had posed no such threat and he had caused no damage. His offending 'must be in the lower range rather than the middle range'.

McHugh pressed on. The offender's conduct was 'in flagrant disregard' of the law. He started to go through the facts of entry again.

Reeves interrupted him: 'He walks up and down a creek bed, he doesn't attempt to go into inner areas.'

'That's a finding Your Honour can make,' conceded McHugh,

before reiterating the requirements of general and specific deterrence. This offending was more serious than Paul's previous trespass on Commonwealth property, for which he'd been fined with no conviction recorded. This was why the Crown said that the 'only appropriate' penalty now was a sentence of imprisonment.

Reeves said he would normally start from a consideration of the seriousness of the offending, and there was nothing about this offending, prima facie, that would attract a period of imprisonment. He was 'struggling' to see why a significant fine would not do.

McHugh's submissions in writing contained this sentence, which he hadn't actually uttered in court: 'The Crown submits that this is a serious offence, potentially striking at the height [sic] of national security, although the offender was caught before any such harm could occur.'

That's where Paul began his rebuttal: 'With all respect, I was carrying a rattle and flowers, walking up and down a dry river bed!'

He contested the characterisation of his conduct as planned, calculated and deliberate: he had been drawn to answer a call and had conducted himself with respect towards all involved.

The Crown's written submissions had referred to 'a number of resources' used to 'detect, locate and arrest the offender and which had likely had a sobering effect on security in the facility'. Paul said the resources were already in operation at the base, and that's where he was arrested after all (it wasn't as if he'd gone into hiding).

Likewise, the Crown submitted that 'significant resources' had been required to prove the charge which the offender contested. Paul countered that he had signed agreed facts and in no way obstructed the course of justice.

But you pleaded 'not guilty', Reeves reminded him.

I had no other option, said Paul, out of respect for 'the integrity of my beliefs'.

There was an exchange about his family and material circumstances, including his ability to pay a fine. At that stage, his older daughter, twenty-two and a student midwife, was living independently, but he was still providing food and board to his younger daughter, nineteen and studying childcare. He had quit his job as a youth worker in order to attend court but he lived cheaply in a shared house, grew a lot of his own food and was confident that he would get work back in Cairns. He would be able to pay a fine, in instalments. He noted the fines received by the Christians Against All Terrorism. The prosecutor had said their case was not similar: 'I say mine is less serious,' said Paul.

He was nervous about the prospect of going to gaol for a considerable time – it would be excessive considering the nature of his acts and his conduct in court – but if a prison sentence was what His Honour considered appropriate, Paul asked for it to be suspended.

Reeves had a final query for the Crown. It was about the offer that McHugh said had been made to have the case dealt with in the lower court, which Paul had been unaware of.

McHugh, however, was able to refer to a file note and the date of the hearing when the offer was made. Whether the information ever got to the offender was a matter he couldn't assist with.

If the Crown had been 'cavalier' in its dealing with the possible 'bias' of Reeves, this too was cavalier. Had the Pilgrims' trespass been dealt with in the lower court, the maximum penalty they would have faced was six months in prison. You would think that this would have been made abundantly clear to them. They were not in court on the day it had been discussed (excused from appearing because they lived interstate) and neither unfortunately was their legal advisor Russell Goldflam; a colleague had appeared in his stead.

When Russell went back over the files, he found a note his colleague had made about the matter which he had obviously overlooked, but everything that had happened to that point made it look like a Supreme Court trial was the course the Pilgrims were on: the Crown had proceeded with charges requiring the personal consent of the attorney-general, a procedure normally required only for extremely serious offences; when the charges were thrown out in the lower court by Judge Trigg, the Crown went to the trouble of re-laying them; the Crown subsequently served a notice on each of the defendants that a committal would take place, which is a preliminary step in a Supreme Court trial; the only previous prosecution under the Act, the trial of the Christians Against All Terrorism, had gone to the Supreme Court; they had been vigorously prosecuted with, at one stage, nine lawyers in court representing various Commonwealth entities; and the Crown had appealed against the leniency of the sentence.

There were mixed reactions to my report of the Crown's call for Paul to serve time in gaol.

'Under a government which imprisons any unjustly, the true place for a just man is also a prison,' wrote Charlie Carter, his real name, quoting Henry David Thoreau. Carter is well known in Alice Springs, a frequent left-leaning commenter on social and environmental issues.

'Rules are Rules' was brief: 'Respect the Rule of the Land – or Pay the Penalty.' 'Horton' attempted to turn the comment back on him: 'You are correct @ Rules are Rules. There is only one law and the Americans need to respect that they are operating in Australia where we have respect for the natural order of things. Any jail time or significant fine for this human pilgrim will bring thunder and lightning from the gods and damage ANZUS.'

John Bell suggested staging 'an annual grudge match between the Pine Gap Patriots and the Pine Gap Protesters … five-a-side. Resolve the Anti-Yank, Pro-Yank hostilities on the footy field. Perfect. The Aussie Way.' 'Peter' replied: 'Not wanting a nuclear target down the road doesn't make me anti Yank.'

'Fred the Philistine' kept up his surprising (to me, coming from him) support of Paul's action: 'If he goes to jail it's going to cost $2000 a week to keep him, for a protest which was for the benefit of all Australians … Let's also remember that America does not allow any foreign facilities on their soil, so why does Australia? If we lived in peace and became a neutral country there would be no need for America's help. Remember, America only helps America.'

John Bell painted a lurid picture of the implications of Fred's position: 'A huge, defenceless land mass, with growing multicultural ethnic unrest, half our assets already sold off to foreign powers and more by the day, China owning our ports, and a chookless-head [sic] rabble loosely referred to as our parliament with no major or minor party [having any] idea how to run a pub raffle, let alone defend our national interests. So, Peace, Bro, get real. It ain't gunna last much longer after we cut ties with Uncle Sam.'

Bell later added: 'The Yanks did a good job of protecting us at Midway.' But 'Fred' wasn't having that: 'We have repaid that debt of Midway many times over with the Vietnam, Korea and the Middle East wars. Lost a lot of good men with no benefit to Australia. Now that they are in Afghanistan they cannot even win that war against people with pitch forks and Massey Ferguson tractors.'

Reeves told Paul he would sentence him before the end of the next week. That was not what transpired, but Paul was intending to stay in Alice Springs anyway, to support his friends through their trial and eventually make his way home with them the way they had come.

NOTHING TO SEE HERE

16 November 2017

When the media contingent arrived for jury selection, we encountered unusual obstruction. The security guard wouldn't let us in, 'following orders'. It was a sign of the heightened atmosphere in the courthouse, with the officious presence of the Commonwealth's representatives and all those AFP officers coming and going as a group, impressive for their sheer physical bulk. After a brief bewildered stand-off at the door, we went in anyway, just as the charges were being read.

'I plead neither guilty nor not guilty,' said Jim, standing there in his bare feet. 'I don't see any reason to plead to a court appointed by a government whose war crimes we are opposing by Pine Gap's existence.'

Reeves took it as a plea of not guilty.

Margaret also refused a straight answer: 'Not guilty of a defence area.'

'I'm sorry, what did you say?' asked Reeves.

'Not guilty of a defence area.'

'You're referring to the area referred to as the Joint Defence Facility Pine Gap area, are you?'

'Yes.'

'Thank you. I will take that as a plea of not guilty of the charge.'

Their answers contained the seeds of what they would attempt to do, more concertedly than Paul had: persuade judge and jury, and whoever else might have been listening, that Pine Gap was being used for aggression, not simply defence. They had acted to prevent this, what they believed to be an 'extraordinary emergency', in order to defend themselves and others. The hurdles they faced in producing evidence were immense, not least because of the diffuse nature of what Pine Gap does and the secrecy around it. They were also thinking like ordinary citizens who have sought to inform themselves, not as trained lawyers. But then juries are made up of ordinary citizens too, which meant the Crown would take no chances.

Seventeen people sought to be excused from serving on the jury. A woman had retired last year from the base after working there for twenty-three years: the gallery laughed, and she was excused. A man was a current employee; three women knew witnesses, security staff at the base. Another man was booked to go on a holiday but said he'd spent thirteen years in the military and his wife worked at the base: 'I'll just put that out there.' More laughter. They were all excused, but Reeves was worried about how small the muster was looking, given the potential for challenges and stand-asides. The five Pilgrims had a right to six challenges each, and the Crown had up to twelve.

For people with reasons unconnected to Pine Gap, Reeves gave provisional leave. Some in the remaining muster had served on Paul Christie's jury. Normally the Crown would challenge on that basis; McHugh suggested that they only be called upon if a problem of numbers eventuated. Their names were removed from the ballot, but again it was provisional. That left only twenty-seven people from whom to get a jury plus a reserve. A woman audibly groaned as the barrel containing the numbered balls was rolled: 'We've got a fifty-fifty chance.'

Margaret challenged the very first person to be called, a woman. I couldn't see any clue as to why. Andy challenged a woman whom McHugh had told to stand aside from Paul's jury. 'I think they don't like women with big bottoms,' she muttered as she returned to her seat which happened to be next to mine. A woman whom Paul had challenged was accepted, which goes to show how randomly subjective the whole process is. The Crown made one stand aside and so it went. By the seventeenth ball they had a jury, by the nineteenth, a reserve. There were two men, and the reserve was a man. I knew some of the women. The Pilgrims were lucky to have them, two in particular: they would be attentive and thoughtful.

Reeves gave them his usual careful instructions about their role. When they were sent out for the morning break, McHugh foreshadowed an application to close the court for part of the evidence – viewing the CCTV footage of the trespass. It was a sign of how tightly the Commonwealth was seeking to control the proceedings, though it eventually looked as much like theatrics.

Section 31 of the *Defence (Special Undertakings) Act* provided for hearings in closed court and suppression orders if the trial judge were 'satisfied that such a course is expedient in the interests of the defence of the Commonwealth'. Counsel from the Australian Government Solicitor would make the application, not McHugh.

Reeves didn't like this: 'You're the counsel for the Crown ... you should be making the application, shouldn't you? It's a matter of evidence.'

McHugh accepted that he could do it but he might need to call evidence 'from those behind me', referring to the Commonwealth's representatives, the same man and woman who had sat there during Paul's trial.

The Pilgrims opposed it, of course. Margaret was more often on her feet than the others: 'It's another way that the Crown is trying to create secrecy around this case,' she said, adding that they all had copies of the CCTV, they'd all seen it.

So had quite a few other people. The security around providing the defendants with this footage, about which the Commonwealth would now make such fuss, had been notably slack. A DVD copy had been sent to Russell Goldflam in March, innocuously labelled, in an unmarked plastic sleeve, with no accompanying request or instruction that it be treated as sensitive in any way.

In court the ABC's journalist asked to speak: the ABC had concerns 'about the prospects of parts of this trial being heard *in camera* [closed court]'.

Reeves said he would hear from him when the time came.

~

Soon there was another wrinkle to iron out in the absence of the jury. Margaret wanted to give an opening address, providing 'a very quick understanding' of why the Pilgrims had done what they'd done, 'context to the prosecution's evidence' which she saw as 'very flat and material'. McHugh was opposed. It turned on the same issue as in Paul's trial. The Crown did not want the legal defences of necessity and defence of another to go to the jury, and they were what Margaret's address outlined.

Reeves ruled against it: not only might the material ultimately be inadmissible as evidence, it was also argumentative, containing 'a series of highly contestable political statements'.

The jury was brought in for McHugh's quite lengthy overview of the Crown's case, as 'flat and material' as Margaret had suggested. It included the facts admitted by the accused, which the jury would see in due course. The prosecution witnesses they would hear in the meantime would 'clarify' these admitted facts, which McHugh said was a 'matter of fairness' to the accused.

During this time I noticed a third person, a man, join the Commonwealth's representatives sitting behind McHugh. The reason for him being there would become clear shortly.

Sergeant Gadsby returned to the stand. He'd been on night duty as the shift supervisor when the Pilgrims were arrested, and it was he who detected on CCTV their approach towards the base. His evidence was similar to what he had given in Paul's case up to the actual trespass event.

At about 3.40 am Gadsby was sure that 'five persons of interest were moving pretty swiftly towards the facility'. They were carrying backpacks or, one of them, 'something': 'We weren't quite sure what it was at the time' (likely one of the instruments). The response team was sent out at 4.07, when the intruders 'breached the facilities'. By 4.22 they were climbing Mariner's Hill – 'a very steep and rocky hill, quite dangerous to climb on', according to Gadsby.

This was not taken up in court, but almost an hour and a half had elapsed since one of his team had first detected, at 2.40 am, an 'irregular movement' on CCTV. A half-hour later, at 3.10 am, Gadsby had been certain enough of what he was seeing to call his boss, Ken Napier. The team wasn't sent out till 4.07 am, almost a full hour later. Why hadn't they gone out earlier and prevented the incursion, as they had done on the Pilgrims' first attempt? When Napier took the stand, he would provide something of an explanation. (Happily the Pilgrims were not intruders with 'malice aforethought', to use an expression of McHugh's.)

For the investigation, Gadsby had requested a copy of the CCTV footage, from the time their cameras had first picked up the approach of the Pilgrims; he used it as a guide for plotting GPS points on a map – as requested by Federal Agent Davey – to show where the arrests were made and where the northern boundary had been 'breached'.

The Crown hadn't quite finished with his testimony, but McHugh said a legal matter 'may need to be discussed'.

'Ladies and gentlemen, would you please retire to the jury room,' Reeves told them.

They were going to miss a fine bit of grandstanding by the Commonwealth.

McHugh was on the back foot. Reeves had already indicated his disapproval of him not taking full charge of any application that might be made. Now McHugh hedged around the matter.

'Your Honour, the Crown wishes to adduce some evidence of the CCTV footage which is of a particular type and to which there is a concern by the Commonwealth, and they wish to make an application, which I understand was foreshadowed at some point, in respect of some matters to do with the defence of Australia.'

'I thought I made it fairly clear before,' replied Reeves. 'This is a criminal trial. You're the prosecutor. You're the person who is able to call evidence and decide whether you should make an application to have evidence dealt with in a particular way. I don't see what status the Commonwealth has in this matter apart from it being the relevant prosecution authority. Do you want to, as the prosecutor, have this evidence taken in a particular way?'

'I do, Your Honour. I'm certainly instructed to do so.'

'You're the prosecutor whether you're instructed to or not.'

McHugh well understood, he could make the application, he was 'in effect happy to do it' but because it was 'an area where there can be some concern', Reeves would be assisted 'in hearing from Mr Begbie', often counsel for the Commonwealth 'in this sort of area'.

It soon became clear that the man who had taken his seat alongside the two other Commonwealth representatives was Timothy Begbie, senior general counsel from the Australian Government Solicitor. A few tortuous sentences later McHugh appeared to propose having Begbie appear as his junior. Reeves accepted this, provided it was 'clearly understood' that the application was being made by the Commonwealth DPP.

McHugh thought that was appropriate, but Begbie certainly did not. He now got to his feet. There was a mirthful murmur among the Pilgrims and their supporters: Begbie was all too familiar to them from his interventions in the Christians Against All Terrorism trial. A message was sent to Russell Goldflam to come into court if he could.

Begbie was 'sorry to interrupt' but perhaps he could explain 'by reference to some authorities and cases' why the application would be 'separate from the advancement of the proceeding itself'.

'I should indicate at the outset,' he said, 'that I do that on instructions from the Secretary of Defence and the commissioner of the Australian Federal Police on behalf of the Commonwealth. So not, as it were, on behalf of the Crown as prosecutor in this matter.'

Reeves explained that such an application 'may open up a Pandora's box'.

Begbie said he understood but went on blithely, coming forward to avail himself of a microphone at the bar table. He was nothing if not verbose, although that is a hallmark of the profession, bulwarking every simple proposition with a few others. It boiled down to wanting to close the court 'to protect the interests of defence and national security' and it being 'conventional and appropriate' for the executive to be heard on such a 'protective application' separately from the Crown.

Numerous propositions later we learned what it was that they were so worried about: that the CCTV footage would reveal the position of recording equipment and its coverage.

We already knew from Gadsby's testimony that he'd been watching the Pilgrims draw nearer the northern boundary for more than an hour, and that he couldn't see everything that transpired on Mariner's Hill:

'You continued to monitor the events occurring at that location through the CCTV, is that right?' McHugh had asked him.

'Yes, as best, as best we could because of where the arrest was taking place on the hill, quite a difficult angle ...'

'But you weren't able to observe, the camera did not have a direct view of some of this, is that right?'

'Yes, that's correct. We could not see the arrest itself.'

Reeves, though, was preoccupied with the principle of not allowing external intervention in a criminal trial. Begbie had said they were 'firstly' relying on section 31 of the *DSU Act*. Given that, Reeves had 'some difficulty seeing why I should entertain the Commonwealth or anyone else separately from the DPP, who's doing the prosecuting of the offence under that Act'.

There were other provisions, Begbie said, in the *Crimes Act* and *Criminal Code*, as well as multiple 'authorities' that he had copies of ready to hand up – all of this not 'remotely intended to be an exhaustive identification' of his proposed course of action. Affidavits from the assistant commissioner of the Australian Federal Police and the Acting Associate Secretary in the Department of Defence would establish that the orders being sought were in the interest of the Commonwealth.

Reeves warned McHugh again that the matter could delay the trial, even derail it all together.

'I understand that, Your Honour, and I understand that there is then a forensic decision that the Crown must make.'

'*You* must make. You're the prosecutor,' insisted Reeves.

Begbie again jumped in, wanting to reply to 'what has fallen from Your Honour'. He wouldn't take long in making the application, 'no more than a further fifteen minutes'. By now, he was really trying Reeves's patience.

'Mr Begbie, you must know that one of the basic rules in criminal trials is that you don't allow interlocutory proceedings to derail the trials. Except in very, very limited circumstances, a party to a criminal trial can't seek to appeal a matter during the progress of the trial. But if your intervention results in an order, that may open up a potential for an application being made elsewhere.' The example he gave was of 'some media representative wanting to agitate an interest in the issue'.

Begbie, however, could not be faulted on persistence. At the risk of 'stretching an indulgence' he felt 'obliged to say' that Reeves refusing to hear an application 'on the basis of that kind of risk' – possible derailment of the trial – would 'involve error'. Again he had references to principles in previous cases.

Reeves said he would adjourn to read Begbie's material, but now Margaret asked to be heard.

'So I'd just like to welcome back Mr Begbie to the court in Alice Springs,' she said, with a big smile to him, before asking Reeves what the Pilgrims' position was in terms of opposing the Commonwealth's application.

If he did 'open the door' on it, he would obviously hear from them, Reeves said.

'It's just that we might have to put some things into place in order for us to sound coherent,' Margaret said. The Pilgrims had a plan for this eventuality, which Reeves didn't doubt.

Reeves wasn't terribly impressed with Begbie's authorities. Only one dealt with a situation similar to this trial, and even then, in his understanding, the applications had not happened during the trial itself.

He suggested that McHugh go on with other evidence and, if necessary, come back to Gadsby's video evidence when the issue of the Commonwealth's intervention was resolved. He would deal with it outside of the trial, early one morning or at the end of one day, respecting the fundamental rule – 'except in very, very exceptional circumstances, and this is not one of them' – that 'no-one can intervene in a criminal trial'.

McHugh pointed out that Russell Goldflam had come into court – he was standing at the back: 'It may be the accused can be properly represented, Your Honour.'

Margaret rose and sought leave for Mr O'Gorman SC from Brisbane to respond to this matter at an appropriate time. Reeves agreed: she could see whether Mr O'Gorman was available and a time would be fixed for next week.

Five days had been set aside for this trial. It was still only day one, a Thursday. The intervening weekend would be useful for the preparations.

The willingness of Daniel O'Gorman SC to appear was another instance of pro bono support for the peace activists from prominent counsel. Back in May, Rowena Orr – Ron Merkel's junior for the appeal by the Christians Against All Terrorism, now a nationally prominent silk herself – had provided them with preliminary advice. They had asked her to consider the effect of the 2009 amendments to the *DSU Act* and whether there was any reasonably arguable defence to their charges of the sort that had led to the acquittal of Bryan and the others. (In extreme summary, the answer was no; in her view the amendments had 'nullified' the effect of the appeal.) As it would work out, there would be no need for O'Gorman to appear but he remained available to advise them to the end.

The jury returned and so did Gadsby, to identify the copy of the CCTV, but it was not tendered. It never would be. If there was any substance behind the Commonwealth's belated vigilance over this footage, we would not learn of it.

Andy started the cross-examination. With his questions he tried to probe the security service's understanding of who the protesters were and why they had concerns about Pine Gap. This seemed to be about getting the prosecution's witnesses to do some of the groundwork for the Pilgrims' case.

'As part of your briefing were you told how many people or what kind of people, groups, would be coming as part of the IPAN conference?'

'Yes,' answered Gadsby. But he wouldn't take it further, they were simply 'various groups'.

Was it mentioned what their concerns were regarding Pine Gap?

Yes.

'Can you tell us, just so the court knows, what some of those concerns were?'

'The concerns were what the Joint Defence Facility at Pine Gap, what we do, what happens out there, yes.'

'What part of the operations of the Joint Defence Facility at Pine Gap were they concerned about?'

'I'm not going to answer that, due to security classification levels.'

Andy tried to press him on this: What had he heard people saying?

McHugh objected, hearsay and 'unlikely to assist the jury in their deliberations'.

Andy said he would rephrase, but after a false start, he handed over to Jim, who asked Gadsby if he was aware that members of the International Campaign for the Abolition of Nuclear Weapons were in Alice Springs at the time of the IPAN conference. Gadsby didn't recall. Was he aware that that group had won the Nobel Peace Prize?

'I recall something but I needed to do a job,' Gadsby replied.

'So are you aware that Pine Gap is intimately involved in any potential nuclear war?' Jim tried.

Poor Gadsby. I say it not because he was unduly suffering, but these questions about what Pine Gap does should be answered by people in leadership positions, not be left for minions to field. The Pilgrims had tried, without success, to get discovery of documents from the AFP, the Department of Defence, the Australian Signals Directorate and the deputy head of the base. They had hoped these would show – as an affidavit from Margaret put it – that the base was linked to a 'decentralised network of military operations which resulted in airstrikes by drones and other military aircraft during the week of our action'.

Although the Crown was using national security legislation to prosecute, under the authority of the attorney-general, they were conducting the case like a simple trespass offence. They were having it both ways. This left the Pilgrims trying to establish through the backdoor of security personnel the facts of the escalating emergency that they believed the base was responsible for.

McHugh objected to Jim's question, on relevance. Jim responded: it was relevant to why they were at that base, to defend others and themselves and prevent an imminent disaster – the whole basis of their defence.

'But how is this man's view about whether Pine Gap might be involved in a nuclear war relevant to that?' asked Reeves.

Pine Gap is part of the US nuclear war machine, said Jim, and 'information supplied by Pine Gap' would be involved in 'any potential nuclear war'. That was important to the Pilgrims' 'whole argument'.

Reeves ruled the question irrelevant.

~

The Pilgrims were buoyed that night by Greens senator Andrew Bartlett drawing the attention of the Senate to 'something that's happening right now in Alice Springs involving a number of people from my home state of Queensland'. He outlined the background to the trial, the peace movement's concerns about Pine Gap and its role in the 'global war machine'. The Pilgrims' action had been a faith-based and clearly nonviolent protest, yet 'the Commonwealth is explicitly seeking to jail these people'.

He read onto the record an open letter to Attorney-General Brandis that had been published the previous weekend in *The Saturday Paper*, under the signatures of seventy Australians, many of them prominent. The letter said in part: 'This prosecution occurs as Australia prepares to serve on the UN Human Rights Council and when UN Rapporteurs have criticised policies, laws and actions of your government that undermine freedom of expression, freedom of assembly and the right to protest. These are fundamental civil rights, and they are profoundly important when governments are engaged in the sort of conduct which Pine Gap facilitates.' The signatories called on Brandis to exercise his discretion to have 'this punitive, disproportionate and expensive prosecution' discontinued before the matter came to court.

That intervention clearly had not occurred.

17 November 2017

Straight away Ken Napier, the base's head of security, cut a different figure from his AFP colleagues. He looked directly at his questioner, speaking in an everyday conversational style, a smile never far from his lips.

He had worked at the base for almost twenty-five years. During the IPAN conference and convergence protests he'd been the commander of Operation Fryatt. Two months' planning had gone into it, with the actual operation lasting a week. In quieter times he was responsible for day-to-day operations of the AFP in Alice Springs, everything from its financial and administrative management to its representation in the local community.

McHugh asked him the well-worn questions about the base's fences. Then it got more interesting.

'Now, do you have any interactions with our traditional owners?'

'Yes, I do.'

'You speak a traditional language yourself, is that right?'

'Yes, yes, I do.'

'Which language is that?'

'Primarily Pitjantjatjara.'

Napier had grown up on a cattle station in South Australia and had gone to school in Oodnadatta, where he was one of only three non-Aboriginal students: 'So you basically had to learn to speak, otherwise you couldn't interact.' Aboriginal people also lived on the station, and when he was sent to do his schooling in Alice Springs, his 'best friends back in the day and still current are Indigenous'. He said he was 'well aware' of the traditions of Arrernte people (the traditional owners of the Alice area).

'And do you have any interactions with those traditional owners from time to time?'

'Over time, yes.'

McHugh asked whether he could discuss what that involved.

'We've had members throughout my employment period who are Indigenous. We actively try to recruit Indigenous locally. It's something that we've gone out of our way to do. But from the operations perspective, we actually have sacred sites within Pine Gap, and we take great pride, especially during my period because I fully understand the sensitivities and how important it is, and we well and truly protect those sites. They're not very well known to even many long-term facility employees except for us. And over time we've facilitated groups of Aboriginal elders that come out to do inspections of the sacred sites.'

McHugh then asked if he was familiar with the name Peter Wallace in that respect.

'Not in relation to an elder that had dealings at Pine Gap, no,' said Napier.

McHugh did not draw a point from this exchange, not now, and not later in his closing address. I drew from it three things (apart from Napier being an interesting man, no cardboard cutout of an AFP officer): one, that traditional owners had a relationship with authorities at Pine Gap, where, thanks to Napier, they were possibly treated more respectfully than in many another local contexts; two, that this would undermine in the minds of the jury the moral appeal that McHugh fully expected the Pilgrims to make, that they had the permission of traditional owners to go onto the base; and three, given his local knowledge, it was surprising that Napier didn't know the name Peter Wallace, who is a public figure and known for his traditional association with the Pine Gap area, putting it like this: 'The sacred site near Pine Gap, that's my mother's and grandfather's. I've seen people standing there, like soldiers, you know … but that's my mother and grandfather's country.'

McHugh returned to details of the incursion. Napier referred to 'replicating' the procedures they had followed during the Pilgrims' earlier attempted trespass. On that first occasion, as they were intercepted outside Commonwealth land, it had been a matter of assisting the Northern Territory police, whom the private landowner had contacted.

On the second occasion, the Pilgrims were 'moving a lot faster and in a straight line through the scrub towards our location' compared to the first time when it had been 'walk, stop, walk, stop, confer' (and finally stop altogether, lay out blankets and have a sleep!).

The Pilgrims' greater speed second time round did not seem to have produced a response in kind. Napier and some of his senior staff met to discuss their 'tactics and techniques'. By the time they had a plan – going offsite to meet the group – Gadsby came and told them that the five were already inside.

They raced to their cars then and drove out to Mariner's Hill. They had in mind 'a pincer movement', sending some members to the front, keeping some in the rear in case the Pilgrims 'decided to run': 'Knowing the terrain was quite rough, it was something that we wanted to make sure we had enough resources for ... We drove to the top of the hill and we had members go to the bottom of the hill and the assumption was that they were in the middle.'

Having to make that 'assumption', we now understood, was due to the blind spot in the CCTV coverage.

Napier with his colleague started to climb down, towards the 'musical noise' the Pilgrims were making: 'We could hear as opposed to see the people coming towards us.' Eventually he was in a position to shine his portable floodlight on the Pilgrims. He registered the presence of the four men, including Jim whom he recognised. He gave an order for Andy to be stopped from filming, but he was more concerned about Margaret: 'I asked her to stop, and then at the time all around me the apprehensions were taking place. I assisted Margaret up over the rock because we'd well and truly established that we were in a fairly precarious position, so – and then down onto the road where she was taken away from me.'

Margaret took the initial cross-examination.

'Hi, Ken.'

'Good morning, Margaret.'

Not many cross-examinations begin this way. Margaret consistently resisted the formality of the court, and Napier had no problem with that.

'Am I known to you?'

'Yes, you are.'

'When did I become known to you?'

'It would be about the time I first met your husband. Probably a decade ago.'

'I heard that you sent condolences through Mr Goldflam after he died?'

'Yes, that's correct,' he answered.

Napier had been part of the response team arresting Bryan during the 2005 trespass and there were many further contacts during the protracted legal proceedings that followed. In the trial he was a witness, but it went further than that.

'Remember Ken?' Donna Mulhearn wrote on their blog during the appeal in 2008. 'The Head of security at Pine Gap who we have come to know quite well in the last two years? He has taken holidays so he could be present at the appeal as an observer. With him and so many other familiar faces in the courtroom one of the DPP lawyers commented "it's like a family re-union!"'

It has to be one of the more unusual bonds between people – a shared history of arrests and trials from opposite sides of the fence. It gave me the headline for my next article: 'Strange encounters: the Peace Pilgrim and the Police Sergeant'. Among the commenters, peace activist and former West Australian senator Jo Vallentine picked up on the 'evidence of respect, and a degree of trust' between Margaret and Ken. She went on to thank the five for their 'strong and spirited witness for peace … Good on you all. Hope this doesn't result in prison time!'

Margaret's point with her opening questions was to establish the lack of threat that the Pilgrims presented to the security of the base, which made their prosecution under the *DSU Act* look so heavy-handed.

What words would Napier use to describe their community? she asked him.

'Driven,' was his succinct reply.

Margaret quite liked that: 'Driven? Driven is a good word. And what's your impression of our integrity as a group of people?'

Napier looked for a way to answer. His impression of the group was formed by his experience of arresting them, 'and I would put that there in criminal'.

'Integrity can … have a criminal element, a legal element, and it can also have a moral element, wouldn't you say, that integrity can be divided that way?'

'In that case, ma'am, if integrity is measured in the ability for us to have a civil conversation, whether it be in the back of the courtroom or in the street, very much so.'

'Would you say we held our beliefs sincerely?'

'Yes.'

'And that we are ... genuine and reasonable people, if a little extreme, but reasonable?'

'When found outside the facility, yes. Inside the facility, no.'

Margaret went on to 'a couple of boring things', including questions about the visibility of the trespass signs at night. Napier agreed that the signs are not illuminated and could reasonably be missed in the dark. Margaret didn't try to make the point stick: 'So that's just a little – you know, we're not really too hung up about that.'

She returned to the nature of the threat the Pilgrims were seen to present. She assumed that a risk assessment had been undertaken as part of Operation Fryatt. What were they measuring in terms of risk?

The threat assessments are developed by the National Threat Assessment Centre within the attorney-general's department, said Napier. The AFP feeds in their information but do not gauge the risk themselves.

Margaret asked what scale of threat they thought they were dealing with during the convergence.

'Protest action ranged in this case from the potential for incursion, whether it be singular or multiple. Also the potential for the use of lock-on devices through to peaceful protests that were identified.'

'Is that the order of risk, high is incursion, lock-on devices is medium, and peaceful protesters at the bottom?'

'It's not measured in that way, it's measured against violent protest. So in this case we take all identified risks equally and we try to ensure that we have the appropriate response to cover all those risks, primarily because we're not provided with an order of the day of how these actions will take place.'

'Was there any violent protest?'

'No.'

'Did there even look like there might be some violent protest?'

'To the direct impression, no.'

'And us and our group, who you encountered two days earlier, have you ever known us to espouse violence in any way or be violent in any way?'

'No.'

'When you saw us on the video you could see five people moving through the bush. Did you suspect who we were?'

'No.'

'We could have been anybody?'

'Absolutely. Sorry, yes.'

Margaret wanted to get to the impact of the Pilgrims' trespass: had it disrupted the activities of the base? This would go, in their view, to the action being more than symbolic, but rather a material intervention in the extraordinary emergency underway as a result of Pine Gap's activities.

She asked Napier about previous incursions, getting to the action in 2005 by Bryan, Jim, Donna and Adele.

'Was it disruptive at all?'

'To?'

'To the base?'

'No.'

'Did the base get shut down?'

'No.'

Margaret put to Napier that an AFP officer in the first trial, during a pre-trial hearing, had said that the base was shut down for five to six hours.

'The front gate was shut, yes,' said Napier. 'Continuity of business mission was maintained.'

'So in terms of a risk assessment threat, maintenance of business must be up there as one of the goals for risk assessment, is it?'

'Ma'am, our role is to ensure that we maintain the security and integrity of the facility. The AFP's role is not to maintain the mission, it is to maintain the security, so we do not go delving into the mission. So in all our risk assessments – keep people out, stop the committing of crime, that's our business.'

Through all this Napier maintained his equable demeanour. Margaret went back to the incursion itself, and CCTV monitoring: 'You checked the video and you saw us walk into the bush. Can you tell us about where the sensors from that video were coming from and where you were looking from to see that video? Because it was not looking at us in the Pine Gap facility, was it? It was looking at us out in Farmer Brown's paddock.'

Objection from McHugh: 'Your Honour, the location of the camera in this facility, I think I can say is sensitive, and it's not relevant, Your Honour.'

Margaret withdrew, switching tack to the music that she and Franz were playing at the time of their arrest. What sort of music was it?

'The guitar and, I couldn't distinguish, but it sounded similar to a violin.'

'Was it like a jig, or was it like rock and roll, or was it like, you know, sing-songy protest music?'

Napier smiled: 'I will have to declare that I'm probably one of the most tone-deaf people who could give a good answer to your style of music, but look, at that time it's a noise to me, and I needed to find where that noise was, so.'

The gallery laughed and Margaret did too, apologising that 'our lament wasn't as good as it could have been, because we were still moving through the bush'.

Both Tim and Jim were interested in what Napier could tell them about the base's relationship with traditional owners. When they want to access sacred sites, a submission is made via the Central Land Council, Napier explained, and it is considered by the Department of Defence. A permit is granted and 'we would meet with them and escort them to those sites. And, we've done that many times.' Even when the visit doesn't involve entry to the base, when 'they come out to some sacred sites right next door and some paintings next door, we go out and meet with them'.

'What process have you employed to ensure that you have obtained the correct opinion or thoughts of the Indigenous people?' Tim asked.

'We don't go through that process,' Napier said, that was down to the Department of Defence.

Jim took this line of questioning further.

'So are you aware of any contract, treaty, agreement signed between the Arrernte people and the government or the defence department or the US government to give access to that land or sell that land?'

'No.'

'Or pass that land on to any of those groups?'

'No.'

After more questions about Napier's understanding of 'why some people might think that the land still belongs to the Arrernte people' and about whether he knew Peter Wallace, as a 'traditional caretaker' for the Pine Gap area, Jim asked whether Napier had known Pat Hayes, a previous traditional caretaker, now deceased.

'I'm aware of Pat Hayes ... but not aware of his role as a – how you described it – in relation to Pine Gap.'

Again, this seemed surprising, especially in relation to Patrick Hayes, who is named on signage at Kweyernpe, the sacred site complex right next to the base that Napier had spoken about.

Jim left it there, moving on to even more contentious ground, asking Napier to confirm that the Raytheon Corporation was the largest private defence contractor at the base.

McHugh objected on relevance.

It was relevant to their defences, Jim argued. The base is involved in providing surveillance for targeting and Raytheon makes 'most of the targeting equipment' as well as 'cluster bombs, fuel explosives, cruise missiles'.

Reeves stopped him there: 'Mr McHugh?'

McHugh again went through his arguments against the jury hearing evidence they might be told to put aside later.

Reeves said he would allow the question but would be stopping Jim after that.

Napier said Raytheon was previously the 'service provider' but no longer.

But they didn't 'take out the garbage or clean the toilets and stuff like that, did they?'

'That's correct.'

'But they are still the largest defence contractor in supplying surveillance equipment and technology for spying around the world, is that correct, at the base?'

Reeves interrupted: 'You can ask questions about Pine Gap and Raytheon Corporation.' He seemed not to have heard Jim's tie-in to the base.

Jim tried again: 'Okay, so are you aware that the Raytheon Corporation is the largest private defence contractor operating at the Pine Gap base?'

'No,' said Napier.

Jim tried next to bring into evidence the actual poster he had been carrying during the action, showing the dead child being picked up from a pile of dead bodies, following a US strike in Basra in 2003. The Crown would tender a photograph of it, but McHugh objected to it being shown to the jury at poster scale: 'In fact, it's a photograph that's been shown around the world, Your Honour. I've certainly seen it before. There is no forensic purpose other than to inflame the emotions of this jury, in our respectful submission, Your Honour.'

Jim argued that excluding it from the court would exclude evidence of what happened at the scene of the Pilgrims' arrest. The poster went to why they were there, what they were on about. It would be 'prejudicial to us to try and exclude it from the case simply because it might help our case'.

'But why do you think it will help your case?' pressed Reeves.

Jim's answer included reference to 'self-defence and reasonable excuse in our reasonable belief that we were trying to prevent war crimes'.

'Well, Mr McHugh, that pretty much raises the defence, doesn't it?' said Reeves.

'It does, Your Honour.'

'I'm going to have to rule on it at some stage,' he mused.

'Your Honour is. And in fairness to the accused, if Your Honour was to allow those defences, then they ought be able to put it to various witnesses.'

None of the accused had yet taken the stand. Reeves still thought he needed to hear their evidence in support of the defences before he could rule: 'If I rule at this stage based on only this photograph, that might do you a disservice,' he told Jim.

So Reeves marked the poster for identification; Jim would not be allowed to put it before the jury now but he would be able to return to it in his testimony.

PSO Ernest Arlott normally worked with the Diplomatic Protection Unit in Canberra. He'd never been to Pine Gap before Operation Fryatt. He was one of the response team sent to Mariner's Hill and his evidence went to what Andy alone was accused of, possessing 'a photographic apparatus in a prohibited area'. This is a lesser offence under the Act than actually taking a photo, attracting a maximum penalty of two years.

When Arlott first saw Andy, he said, he was already on the ground, but he was still holding the phone, raised in the air.

'It was facing towards the Pine Gap area,' said Arlott.

'Can you see Pine Gap from that area?' asked McHugh.

'I would say at the time, no.'

'And that is because you're on the northern side of a hill?'

'That's correct, yes.'

The phone was produced in court, in its 'pinky-peach case' (it belonged to Margaret), and Arlott identified it.

Andy took the cross-examination. His questions, as with Gadsby, all went to what Arlott might have been told in his briefings about the reasons why people were protesting.

'I wasn't really sure what the protesting side of it was going to be … that wasn't discussed with us,' said Arlott.

Andy pursued the same line of questioning with Matthew O'Neill, the officer who had tackled him with a manoeuvre he described as an 'arm-bar takedown'. O'Neill said he'd had many briefings and 'with all due respect' he couldn't remember what the protests were about;

neither could he comment on the music that was playing, he had a thousand other things on his mind.

What interested me in his testimony was again the timing of the intercept. As head of the response team, he had them 'standing to' when he went into the briefing room with Napier at 3.45 am. It wasn't until 'shortly after 4 am', when they 'witnessed' on CCTV five people coming through the northern boundary fence, that he and the rest of the team got into their vehicles and headed out for the intercept.

Margaret had a bit of fun cross-examining Federal Agent Duncan Munro, normally with the Protection Liaison Team in Melbourne. At the time of the trespass he'd been sent to the northern boundary area where he had found the instrument cases that Franz and Margaret had evidently dropped once they started playing the lament.

'How old do I look, honestly?' Margaret asked him.

She was challenged on relevance but said it went to the testimony about the rocky ground. Munro had said the hill was 'quite steep' and 'sharp broken rock' was common. The question was allowed.

'I know how old you are, you look commensurate with your age,' said Munro.

'Do I look fit?'

'In what context?'

'For climbing hills.'

'I believe so.'

'One of your friends described me as an "elderly female",' said Margaret. 'I crossed it out and wrote "direct action goddess".'

At this even Reeves, who could get a bit irritated by Margaret's antics, allowed a smile to escape.

Andy tried his line of questioning on Munro, about the AFP briefings on the nature of the protests.

There seemed to be a variety of issues, answered Munro, 'alien abduction' was one.

With this response, Andy would have felt again his frustration with the conspiracy theories and dope-smoking of some at the Healing Camp in 2016. He tried another tack. Would he have been surprised, he asked Munro, that people were concerned over Pine Gap's role in targeted air strikes and drone bombings?

Reeves ruled the question irrelevant.

Federal Agent Davey, even before cross-examination, shed some light on the kind of briefings they worked with. The term used by the AFP is 'issue-motivated groups', he told McHugh, and his role, as we'd heard during Paul's trial, was about 'building rapport' with them, more than obtaining their group names. Apart from visiting both protest camps, offering to facilitate their lawful protests, he'd been to a workshop in town where posters and placards were being prepared, and to a public screening of a documentary about drones, where he'd had a cup of tea. He didn't take notes, that would have looked 'overly officious'; his conversations were on 'friendly terms, backwards and forwards'. He estimated there were twenty to thirty people camped at Hatt Road, and another thirty to forty at the Ghan Pans.

He'd conducted the interviews with all five Pilgrims after their arrest. Only Margaret's went much beyond a basic confirmation of identity. McHugh proposed to play them all now. Margaret objected 'on the basis of boredom – we really only need to hear mine'. McHugh played three, nonetheless, hers as well as Jim's and Andy's.

20 November 2017

Over the weekend the Crown decided not to pursue showing CCTV footage of the trespass, and so any application to close the court would be unnecessary. All that fuss for nothing. The focus now was on pushing harder against the Pilgrims' hoped-for defences. In the Crown's view, the evidence supporting the defences could be obtained 'through a voir dire' – in front of the judge only – allowing Reeves to make his ruling without exposing the jury to the evidence.

In relation to the 'extraordinary emergency' defence, Reeves would need to hear evidence of what purpose the accused had, said McHugh; and in relation to 'self-defence or defence of another', it would be about what they reasonably believed. The Crown could accept, for present purposes, their belief that the Joint Defence Facility was 'the source of all evil' and part of 'a war machine'.

And what was the Crown's view about the 'so-called expert evidence' that the accused were seeking to put? Reeves asked. The experts in question were Richard Tanter and the former senator Scott Ludlam, who had arrived in Alice Springs and were waiting outside in the lobby.

That evidence would not be required if he ruled against the defences, said McHugh. Although another way of avoiding the difficulty it might present was by the Crown accepting 'that the accused had these beliefs'.

There was still an issue of 'identifying the bounds of the emergency by reference to what Pine Gap is used for', wasn't there? said Reeves. And of identifying the people involved in that emergency.

Even if that evidence was heard in a voir dire and Reeves allowed the legal defences, McHugh replied, there would be no need for the expert evidence because the Crown 'will accept in open court that they had this belief ... that Pine Gap is being used for all sorts of terrible things,' said McHugh, adding that there might be 'a form of words' the parties could agree on for the purposes of the trial.

Then, as often seemed to me to be the case, he tried to have it both ways, suggesting that the proposed evidence would be mere speculation and hearsay: 'It is all third hand and it really doesn't rise up to any height, Your Honour ... it is just so hypothetical.'

That though, countered Reeves, was in part what the expert evidence was directed to, 'removing, to whatever extent is possible, the speculative hearsay'.

'Well, Your Honour won't be able to rule on that,' said McHugh. 'That's what will make a mockery of this court, with respect.' He referred to the demonstrations outside the court, attracting media attention, as part of the accused's strategy to use the case to promote their beliefs. 'It must come close to an abuse of the court's processes, Your Honour.'

McHugh also raised the issue of the *Parliamentary Privileges Act*, which was likely to apply to the evidence given by former senator Ludlam.

How would he know that until he saw what the evidence was? asked Reeves.

Even if the defences were allowed, McHugh replied, the expert evidence 'cannot rise any higher' when the accused's beliefs were 'not the question'.

After all that – and I have presented a highly compressed version – Margaret told Reeves they would be withdrawing the expert witnesses. They still intended to call Ludlam and Tanter, but not as experts as they weren't confident that they knew how to handle the rules around expert testimony. Even so, McHugh would shut down these 'ordinary' witnesses at every turn.

First, though, the Crown had to conclude its case. Davey returned to the stand. In cross-examination Andy asked him about the film screening he had attended as part of his effort to liaise with the protesters. Did he recall what the film was about?

'Basically it was the dealing with the use of drones in a military sense,' said Davey.

Did he speak to people about what their concerns with drones were?

'I think I was told by various people what their concerns were, it's not something that I was canvassing with them.'

Wasn't that his job, and to pass on concerns to authorities?

That was so, said Davey, but it was not really his position to comment. He didn't recall drones being linked to Pine Gap, but assumed that that was the 'intention of screening that documentary'.

Andy asked him about the IPAN conference and its program.

Davey had not attended and professed to know little beyond the fact of it occurring and bringing many peace protesters together; he did not know, for instance, that Ludlam had been a speaker at the conference.

Operation Fryatt had not paid very much attention to the protesters and their concerns – was that 'normal procedure'? asked Andy.

'There is no normal procedure,' said Davey. 'My role is not to delve into the weeds of what's protested about, my role is more to obtain, you know, to establish a rapport. So, my main concerns are what physical activity is going to occur at the protest.'

Andy pressed on, about the attendance of plain-clothes police at the conference and whether they had reported back ('Not to me, no') and on the attendance, as a speaker, of the Australian chair of the International Campaign to Abolish Nuclear Weapons (Richard Tanter). Davey knew nothing of that either.

'There's a lot of people being surveilled here,' said Andy, 'followed by Operation Fryatt but there's not much information about what kind of people, what these people are about. What kind of information is important to the police in this situation?'

'In fact no individuals were surveilled, as you say,' replied Davey, 'and the only information that we are really interested in is what is going to occur at the protest.'

Margaret asked him if she specifically had been 'followed in the surveillance period'.

Davey said her activities had been 'of interest' to them.

She asked if anybody, in their group especially, had been surveilled using covert technology? No, said Davey.

She also asked him about the lament she had played at the drone film screening: was it happy or sad music?

'Well that's subjective,' said Davey. 'It was, I guess, yes, a sort of sad violin type music. I'm not a music person.'

'The police obviously need to do some work in that area,' she said, cheekily.

Jim took the opportunity of the photographic evidence that had been tendered, which included a photo of them on Mariner's Hill praying around his Basra poster, to ask about what the poster showed.

Davey had nothing to contribute on this, except conceding, after prompting by Jim, that it showed 'somebody carrying a baby'.

The Crown's final witness was Damion Millar, an investigator with the AFP, Davey's offsider in the interviews with the Pilgrims and, like him, normally based in Perth. His evidence went chiefly to Andy's live-streaming during the trespass. He had got a colleague to make a video of him, Millar, watching the footage as it appeared on Margaret's Facebook page, to which it had been live-streamed, and on the Close Pine Gap Facebook page to which it had been uploaded. This video-of-a-video was screened in court. Millar gave evidence that Andy's footage showed 'someone or some people playing musical instruments in what appeared to be a dark or blackened area, as in like it was nightfall, on the side of a hill'.

In his cross-examination, Andy once again tried to get an AFP perspective on the events they were sent to control. Had Millar attended the IPAN conference?

Yes.

Had he bought a ticket?

He didn't remember, he didn't think so, no.

'No, yes,' said Andy. 'Some people were unhappy with the police turning up and believing they didn't have to buy a ticket. It costs a lot of money to run events like this, but anyway.'

He asked Millar what the conference theme was.

He couldn't recall.

What stuck out for him?

'A particular Aboriginal lady singing me a song or singing everybody a song, saying that she had been taken from other people. It was quite an extraordinary event and we weren't there that long.'

This answer struck me as particularly disingenuous. In the course of a program packed with serious speakers, including national figures and international guests from Guam and Okinawa (both sites of US military bases), this was the only thing he could remember? Andy pressed on, fruitlessly.

What was the purpose of the police being there? he finally asked.

'We had had some reports from the local police that cars were being broken into in the area,' said Millar. 'We just wanted to make sure, where people were, people that were seen out at the base, if they were in the same location, whether we could get any intelligence of any other incursions into the base. Just things like that. I wasn't there for any other agenda but to ensure the security and safety of people on the base and the security of the operation.'

Margaret's and Jim's questions, about surveillance of the group, and about the poster, got the same treatment: nothing to be seen, or heard, here.

McHugh then tendered the Admissions of Fact, with copies for the jury, and closed the Crown case.

ORDINARY OR EXTRAORDINARY?

20 November 2017

Margaret had picked up a cold over the weekend, so her voice was a bit croaky as she made the opening address, appealing to the jury as 'our neighbours and our peers, as ordinary people'. She wanted to explain why she and her friends were not criminals and why their actions were not criminal acts, but rather 'peaceful efforts to prevent other harms being done'.

Much of their testimony would rely on what was already on the public record – 'because that is all we have access to as ordinary people'. It would go to their belief that an 'extraordinary emergency' was unfolding at Pine Gap, transformed from 'predominantly an intelligence-gathering facility to one that is entwined in aggressive and illegal military operations'. Their actions were 'in defence of others' and would 'create a direct intervention to the killing and destruction'. They had hoped to be effective in the tradition of 'Christians bearing witness and being prepared to stand up for the values that Christ embodies – which is against killing'.

In effect she had outlined the defences that the Crown had said all along should not go to the jury. McHugh was immediately on his feet for legal argument and the jury was sent out for lunch. There was a sufficient evidential basis in what Margaret had said, McHugh suggested, for Reeves to rule on the admissibility of the defences, without any further evidence being given.

If McHugh had previously argued that Reeves should follow Justice Thomas in *R v Law* (the Christians Against All Terrorism case) and rule against the defences, both Margaret and Andy now also argued that Reeves should follow her and hear all the evidence before ruling. McHugh rehearsed all his arguments again on why the jury should not hear the evidence, particularly evidence 'likely to inflame', such as Jim's poster. He suggested that the length of deliberations by the jury in *R v Law*, some five hours, might have been due to their confusion, 'having been exposed to that evidence and then told to ignore it'. (An alternative view, of course, could be that at least some members of the jury were reluctant to convict.)

Reeves decided in favour of the Pilgrims. The 'overriding consideration' was that they had a right to present evidence in their defence and any confusion arising for the jury could be dealt with by directions: 'It has been said often that courts should presume that juries follow those directions and that's the way I have proceeded and intend to proceed.'

McHugh raised one last difficulty, which was that evidence might be reported by the media. That though, now that Reeves thought of it, was another reason why it shouldn't be done by voir dire. Whatever reporting was done would be of evidence the jury had actually heard themselves, and he would direct them on its admissibility.

Scott Ludlam took the stand in a private capacity. 'I'm between jobs at the moment,' he said, in a wry reference to his resignation from the Senate four months earlier, as one of soon many parliamentarians caught up in the dual citizenship crisis. He had a plane to catch the next morning which was why his evidence was being heard before the Pilgrims gave their own.

Andy would take him through it but couldn't help pointing out first that it would be

> a little stilted because there are certain things we can't talk about in the court, due to laws of parliamentary privilege. And so, we're not going to talk about those things, despite that fact that all of us defendants followed closely the work that you did in the parliament,

> questions on notice, work on the laws that bring us here today, the estimates arguments and the response to ministerial statements. Those will not be part of your evidence, despite the fact that they did influence us, a lot.

This was heading straight for objection, possibly legal argument and the jury had only just taken their seats. 'Could you just proceed to ask questions, please?' prompted Reeves.

'Yes, sorry – yes. Just a way of clearing air,' said Andy. So he asked Ludlam about the work he had done on Pine Gap outside parliament. With a large number of like-minded people from WA he had travelled to Alice Springs for the 2002 convergence, camping not far from the base for about a week. It was after that experience that he started researching the base's role: 'I knew that it was important but I didn't really understand how.'

What had he learned about Pine Gap in relation to US drone warfare? Objection. Reeves told Andy to make his question 'direct to the witness'.

Andy tried again, but this is one of the hardest things for a lay person to do in a trial, to get on top of the allowed form of questions and to narrow their focus. The form of the question was 'too wide and so on', said McHugh, and was it really relevant to an issue in this trial?

'Well the nature of the emergency is relevant, is it not?' answered Reeves.

'Putting that to one side …' started McHugh.

'How can you?' asked Reeves.

'It can only be, in our respectful submission, the accused individual's beliefs, not the basis for those beliefs,' said McHugh.

'But the jury, if they are required to assess this matter, will have to, for example, determine whether those beliefs are fanciful,' said Reeves.

But Ludlam could not speak to the subjective beliefs of the defendants, argued McHugh. That would go well beyond the bounds of what the jury would be asked to decide.

What was in question, Andy countered, was the reasonableness of the belief that Pine Gap was used for drone targeting. That's why Mr Ludlam had been called, as somebody 'with some level of expertise on the matter to talk about it'.

Reeves allowed the question.

Ludlam said the base's involvement in the 'drone assassination program', undertaken in at least six countries that he was aware of, had been very widely reported and not disputed by United States authorities. It raised 'moral and ethical questions' as well as 'deep legal questions', and caused him 'grave concerns as an Australian citizen —'

Objection.

'You weren't asked about your concerns, Mr Ludlam, you were asked about what you knew about the role of Pine Gap in the drone program,' said Reeves.

Ludlam tried to explain how he had come across the information; it was 'the work of experts, people from inside these programs in the United States and elsewhere, open source reporting, the work of whistleblowers that provides a reasonably complete case which is not being repudiated by United States authorities or Australian authorities'.

Objection.

Reeves told Andy to move on, so he asked Ludlam about the legal questions involved in the drone program.

Objection.

Without hearing reasons, Reeves ruled that it was irrelevant.

'Well I assumed, since we're in a legal court, questions of law were relevant,' said Andy.

'I don't need Mr Ludlam to tell me what the legal position is, when we come to the end of this trial there will be a debate about that,' said Reeves (although surely there would be no further debate in this trial about the legality of the US drone program).

Andy tried to move on to the role of Pine Gap in a potential nuclear war.

Objection: breach of either parliamentary privilege or hearsay and opinion rules.

Reeves could see how the drone program might relate to their defences, but not the broader threat of nuclear war.

'Many people would consider nuclear war an extraordinary emergency,' tried Andy.

Reeves ruled the question irrelevant. And so it went. Andy's questions fell like dominoes: surveillance, irrelevant; Edward Snowden's leaks, irrelevant; actual harms being done from Pine Gap, irrelevant; the nature of the emergency caused by Pine Gap, hearsay, opinion or breach of parliamentary privilege.

Andy was remarkably resilient in the circumstances. He moved on. In an effort to establish that the Pilgrims' actions were an effective way of trying to stop the emergency, he asked Ludlam about his experience with actions of civil disobedience and his knowledge of how such actions create change.

'My direct experience has been that it's a legitimate form of social change, particularly when it's conducted in a nonviolent way, to bring attention of gross injustices to the wider population, to propagate the ideas of the various campaigns. And I don't think that there would be much argument that, historically – and right up until very recent times – these campaigns have formed a really important part of the way that democracies evolve – and that societies have evolved over time.'

General questions beget general answers, but at least there was no objection.

Andy moved on to the IPAN conference in 2016. Ludlam described the broad subject areas he had spoken to – US drone operations, preparation and deployment of nuclear weapons, global mass surveillance.

Andy asked about the role of such conferences in bringing to public attention knowledge about places like Pine Gap.

Objection.

Andy explained what he was attempting to establish, that the Pilgrims' actions – trying to make better known what happens at Pine Gap – were a way of defending others. Reeves thought he had gone far enough with that.

Margaret wanted to ask some questions. Normally that would not have been allowed (one person only would lead evidence) but

McHugh did not object 'in this instance'. She showed Ludlam a photo. It was of four women and a man, holding a large rainbow peace banner and a big placard saying, 'Quaker Grannies are great.' The women were wearing the bonnets of the Quaker movement, and one of them, Ludlam said, was very familiar to him – his friend and mentor of twenty years, Jo Vallentine: 'For those who don't know, she was the first person elected on a platform of nuclear disarmament to any legislature anywhere in the world, and she maintains a really important role in the peace movement of somebody who has had a foot in the legislature and had a foot in civil society organisations for a really long period of time. She's also a wonderfully inspiring individual.' In the photo he recognised Margaret too, and noted that three of the five were wearing Peace Pilgrim t-shirts.

'Would it surprise you to know that that's at the back of a watch-house on release?' Margaret asked him.

'It would be utterly unsurprising, Ms Pestorius.'

Margaret wanted to tender the photo. The point was, I assumed, to show the Pilgrims in the context of the wider nonviolence movement that on occasion was also involved in civil disobedience – and who more disarmingly respectable than the Quaker Grannies?

'Your Honour, I'll say relevance, but I won't press it hard,' said McHugh. Reeves allowed the tender.

There was no cross-examination – no way was McHugh going to allow any further oxygen for civil disobedience and the like. Ludlam could go.

Reeves reminded the Pilgrims that their individual testimonies should be about what they saw, heard, did or felt in relation to their beliefs, but should not go into their arguments about the right or wrong of their actions: that would be for submissions, later. McHugh observed that they were likely to breach the division, 'no disrespect' to them, but the Crown's real concern was with documentary, audio or visual material the accused proposed to tender.

From the stand, Margaret spoke movingly of her life and values and McHugh did not interrupt. She spoke of being a widow: 'I had an amazing marriage with somebody who was very driven and a very brave person.' Of being a mother to a nineteen-year-old 'who I brought into this world to experience beauty and love – creation'. Of her love of music; of her work as a therapist, healing people from trauma, 'and often that's the trauma of war at some level or another'. She spoke of living in a large communal house with people who 'share a vision of the future, of communities that are safe'. Of being part of the local Catholic church, for twenty years: 'I participate in the theology and the liturgical work of that church and it's a regularity I have with ordinary people, just like my work is with ordinary people.'

Over that same period she had also been in a community practising nonviolence: 'I generally do that in holidays and, you know, our holidays are often exhausting because we go somewhere where we resist the violence of the state.' She'd been doing peacework from as early as 1991, during the first Gulf War. With Bryan in Cairns in 2001 she had set up a peace group, focussing on warship visits to the city: 'We made relationships with people around the world doing this sort of work.'

It involved 'all those ordinary things that ordinary people do': trying to speak to parliamentarians, running small actions in the local community, writing submissions – 'my husband was a prolific writer'. Of course, they had marched with the millions in 2003 against the Iraq War, and continued over the next ten years to try to break into the discussion about the US military and what they were doing in Australia and to Australia, and overseas. So, it wasn't that they hadn't tried other processes.

She spoke of the visits to Cairns by American peace activists, Malachy Killbride and Kathy Kelly, and the growing 'sense of emergency' they imparted, that the situation was becoming 'more complicated and more severe'. Then the pivotal moment came: former prime minister Malcolm Fraser was speaking on the radio, saying Australia needed to close Pine Gap and reappraise the US alliance. Here, suddenly, was this eminent person 'sitting inside our voice'. She had listened over and over to his interviews, given in a seven-month period in the year before he died in 2015.

Attention was being drawn to the drone program from about 2011, Margaret said, but it didn't hit her consciousness until 2014, when whistle-blowers were starting to come out. She referred to the Bureau of Investigative Journalism and their work on drone strikes in Iraq, Afghanistan, Syria, Pakistan, Yemen. The strikes were rising in Afghanistan; having fallen, they were going up again, fifty strikes a month, a hundred strikes a month: 'So, that's maybe three strikes a day, you know, it's going up ... What we've done up until now hasn't worked. Maybe if we call on God, maybe we call on our bodies – the body of Christ, our love – we go to this place and we see what we can do to try and stop what's going on there.'

She wanted to play in court an extract of an interview with Fraser. McHugh was not keen, it was simply buttressing, but Reeves allowed it.

It was extraordinary to hear Fraser, that unmistakeable voice summoning all the authority of his seniority and position, the authority of someone who had reflected deeply and changed, calling for the same thing as the Pilgrims were, as many at the IPAN conference had – for Pine Gap to be closed because it was now involved in targeting for drone strikes in countries with which Australia was not at war; to hear him say that Australians operating out of Pine Gap involved in that targeting had no legal cover and could be vulnerable to charges of crimes against humanity; to hear his belief that drone warfare would eventually be outlawed by international agreement, and his argument for Australia to make its own foreign policy decisions: 'I really think it is totally un-Australian to have a situation in which it is America that is in charge of our destiny.'

McHugh reiterated his objections to 'all this evidence'. Hearsay, 'bootstrapping the uncontested beliefs of the accused'.

Reeves was firm: 'Mr McHugh, I have already told you that I have the view that the Crown's concession – if I can describe it as that – about the evidence relating to the accused's beliefs does not constrain them on the evidence they call in their case. Whether or not it is bootstrapping, or any of the other things you have described it as, is a matter really for the jury, isn't it? It's not a matter of law that I can exclude.'

'Well ...'

'It is not something I can exclude.'

'Well, Your Honour could exclude it ...'

'On risk of this trial going off the rails.'

McHugh said Reeves was right to 'err on the side of caution in ensuring a fair trial' but nonetheless protested that, with the interview having been played, 'the horse has bolted, the gates have been opened and the Crown is left prejudiced in that regard'.

Reeves accepted tender of the CD copy of the interview and, against objection from McHugh, a graph showing the rising incidence of drone strikes in Afghanistan, 'as something that formed our belief of escalating danger'. Margaret also hoped to tender a bundle of articles about Fraser, by Fraser, about Pine Gap but she stumbled on procedure. Reeves asked her to review it all overnight. The jury was sent home.

'Are you standing for some reason, Mr McHugh?' Reeves asked.

'No, Your Honour ... Out of habit, Your Honour.'

McHugh could occasionally be droll.

21 November 2017

The next morning McHugh was back to playing hardball. He wanted to qualify part of Scott Ludlam's evidence, about the base's involvement in the 'drone assassination program' and the 'grave concerns' it caused him: The Crown did not want it used to prove the truth of the facts.

The argument came up again as Margaret completed her testimony. She tried to turn it to the Pilgrims' advantage: 'It's impossible for us to prove those things at this point in time. That's the whole point. And that's why we don't go to public policy to try and get change, because public policy has totally failed us ... And so we went there directly to try and stop the strikes, to try and stop that information going from A to B and from B to C and C down to some poor villager on the ground who is being killed in the middle of nowhere.'

She tendered an article from *The Guardian* about the drone strike on Shadal Bazar on the same date as the Pilgrims had set out on their walk into Pine Gap. McHugh objected. Reeves asked how it was relevant, accepting its tender after her reply:

'It's relevant because you're going to ask us, "Who are the victims?"

I've heard you ask that question already, "Who are the victims here?" These are the victims.'

In cross-examination McHugh set about dismantling the possibility of the Pilgrims' action being seen as reasonable and necessary in terms of their defences. In return Margaret gave him a lesson in nonviolent direct action and Christian activism, whether he wanted it or not.

He asked about Margaret's involvement in things like writing letters to parliamentarians, voting, in short being involved in 'the social polity' to make change. She replied that the action was not about lobbying politicians, which has 'its own realm ... a bit like a courtroom, it's a structured thing that's got ways and means. But I don't think it's very effective and it hasn't, certainly hasn't been effective for the peace movement.'

McHugh asked about her involvement in protests (of the lawful kind), including the ones held outside the court during the trials, associating herself with signs saying 'Close Pine Gap'.

For Margaret that was 'an almost ceremonial liturgical act, standing with friends, praying with visual symbols that communicate where our hearts and minds are ... It's not necessarily a protest. Protest is a very dull and overused term.'

But she wanted to communicate her ideas about Pine Gap to the wider public, wasn't that right?

'Absolutely ... because, Mr McHugh, if we do these things in the darkness we are just part of the problem ... That's what St Paul said, bring things into the light, don't do things in darkness. If we do it in darkness, we're part of the denial that exists around that place. We are trying to break the denial. That's part of the purpose of the action. Not the only part, but it's part of the purpose.'

Had she spoken to the media?

'I've tried to here, but they, you know, they're not very responsive.'

(That was a bit cheeky, I thought, given the time being invested in reporting on these trials by the *Alice Springs News*, the ABC, and *The New York Times*.)

What about her involvement with Facebook pages and the Close Pine Gap website?

> Look, some people do those things. I don't do those things. I do actions in order to embody justice in the world. I see myself as part of a body of Christ trying to bring the Kingdom of God into the world. So when I do an action, I'm about putting my ordinary little body on the line. I am not a big powerful lobbyist. I'm not an academic. I live in Cairns, which is a long way from anywhere. People from Alice Springs will understand this. I'm not close to the centre of power. And even the centre of power in Cairns, which might be Warren Entsch or Rob Pyne, is not the centre of power. They are very marginalised in the parliament. We are a long way from anything. I do not trust in my ability to influence the parliament through anything I am doing. If the parliament responds to a change in values from the people at some point, or if the people force a response on the government, then hallelujah, we will have change. But there are a lot of other steps before that. That is not my focus. That is the ordinary way of seeing power change and I think it's a failure of a way. I am a nonviolent activist and being a nonviolent activist is about disruption, stopping harms, facing up, witnessing, speaking truth to power. It's not about lobbying politicians so that they might change, because they don't.

There was more, but eventually the direction changed as McHugh tried to undermine any sense that the action might have had a reasonable prospect of defending anyone or anything.

Her calling was to lament, that was the central idea, right?

No, it was to disrupt, but first they had to 'weep and lament', then see how they would be 'called'.

But the plan was to go through the boundary at Pine Gap, wasn't it?

'The plan was to lament on the hill close to Pine Gap ...'

And she had expected to be intercepted by the police, right?

'There's been miracles, Mr McHugh, there's been miracles. The last crew that went in, went in.'

On and on it went, with Margaret putting up a spirited resistance and McHugh showing his years of training and skill in seizing on every choice of words or inadvertent comment to undermine her credibility in the eyes of the jury. He was not the most eloquent composer of sentences but he did not let a chance go by. He finally got to this proposition:

'This idea of disruption is a recent invention to fit in with a proposed defence?'

No, Margaret insisted, the lament came as a secondary idea for the action: 'We had just been at Talisman Sabre when we designed it, well, it evolved over time but I can remember the primary idea was disruption. Disruption, disruption, disruption – try and shut the place down because there are deaths.'

McHugh kept on pressing, more and more questions attacking the plausibility of disrupting anything at Pine Gap. Margaret kept up her refusal: 'I saw my husband do things that I never imagined would happen ... I also believe in the transformative power of lament. I also believe in the transformative power of witness. I don't know what is possible.'

It got to the point where she protested about the repeated questioning: 'You're just like trying to corral me into saying the thing that you want me to say, but I don't agree with that.'

Reeves reminded her that it was a cross-examination and she was required to answer the questions.

It was a relief to take a break for morning tea. When Margaret returned to the witness box, a Corrections officer patted her down, one of the illogical routines of the court. She had not been patted down as she sat at the bar table conducting her defence, but now, a couple of metres away, she was.

McHugh returned to his attack on the plausibility of their disruption. In the 2005 trespass people had used boltcutters to get into the base itself, and she didn't have boltcutters or any such intention this time, did she?

'We had young men who could climb,' said Margaret.

Climb the barbed-wire fence?

'Do you want to show us the fence?' retorted Margaret (knowing that it was a sore point). She had also seen a seventy-year-old man climb over the fence in 2006; she had photographs of it.

McHugh moved back to 'strikes' – he never used the word 'drone' or any term for how the strikes might have been delivered. She believed strikes were happening all the time, right?

'I don't know if it's all the time,' said Margaret, 'but there's obviously several a day and that signals are being collected all the time and that those signals are getting moved around all the time.'

'Your belief would be that if it was in defence of another, an interruption would have to coincide with particular signals coming through Pine Gap involving strikes?' put McHugh.

'—Presumably, yes.'

'So you believe this was happening all the time if you were to have any effect of defence of another?'

'That's right.'

'You could even call it ordinary, what happens at Pine Gap in terms of those strikes, if that's your belief?'

Margaret was taken aback by this logic: 'I don't understand what you're saying … Are you saying it's not an extraordinary emergency?'

'That's a submission later,' said McHugh, 'but you understand what I am going to, don't you?'

'No, I don't think so … what I was saying was that those strikes are increasing all the time and Australia's culpability is becoming clearer and clearer … It's escalating. The numbers of antennas there are growing. The number of signals has become incredibly huge … if you want to call that ordinary, then I don't know …'

McHugh had her on a pin: it's happening all the time, that was the implication of their claim to disruption.

'Obviously we can't know about particular strikes,' Margaret tried, 'or the big plan at particular times and particular places. We are not connected in that particular way. That would be almost impossible …'

'You're an intelligent woman, Ms Pestorius, aren't you?' McHugh finally put. 'You understood when I was talking about ordinary that it's not extraordinary, isn't that right?'

Margaret tried to shift the emphasis away from extraordinary emergency to defence of others, but that too was a 'vain hope', McHugh suggested, 'fanciful'. Margaret fell back on miracles:

'Miracles happen, Mr McHugh. Miracles happen and you bring God's power and intervention into the world.'

McHugh was doing a job, I had to remind myself, but that job reflected an entire framework for thinking about the covert military operations conducted out of Pine Gap – nothing to see here, an ordinary day. It was chilling to hear.

The Pilgrims presented such a sharp contrast to the Crown's phalanx. It was all heart-baring, principles and faith on one side, all faceless power, calculation and manoeuvre on the other. Of course McHugh had a face, he had a professional role and obligations, but there was this strong sense in his management of the case of the shots being called from afar, or not so far actually, from just behind him. I asked those Commonwealth representatives (aside from Begbie) who they were, who they worked for. 'The Commonwealth' was as much as they would disclose. One of them was a woman in her thirties with wavy copper-coloured hair – 'Jessica Chastain' one of the journalists dubbed her, from the movie *Zero Dark Thirty*.

Andy was the next to take the stand. He spoke of growing up with war, distant wars that had entered his consciousness as a young man in country New South Wales via images in the media – the Abu Ghraib 'torture camp' in Iraq, Guantanamo Bay. He saw the people affected as connected to him by their 'innate humanity of sharing this earth'. That Australia was involved in this war gave him a 'personal sense of responsibility'.

Pine Gap was one of Australia's key contributions, he said, outlining its various activities and how he had gained his information: articles by investigative journalists like Philip Dorling, who had a whistleblower source inside Pine Gap talking about the way it was used by the US in drone targeting; like Jeremy Scahill, who wrote about the use of metadata to target drone strikes; like the accounts of 'US drone whistleblowers', such as Brandon Bryant, a video gamer, good at killing digital avatars in video games, recruited into the US military 'to essentially do the same thing, except with real people, as a drone console operator in the US'.

He told the story of Zubair, a thirteen-year-old boy from Waziristan, a province of Pakistan, who testified at a US congressional hearing after losing family members and being injured himself in a drone strike. He had said something that Andy found unforgettable: he no longer loved blue skies because when there were blue skies, there were drones overhead watching everything he and his family did. He preferred grey skies.

Andy spoke of concerns with the drone strikes raised by human rights lawyers, 'that this is just summarily executing people without a trial'; of 'the dehumanisation' of this kind of war, with its decentralised steps separating the person triggering the weapon from the person doing the tracking – 'a very worrying step for humanity'.

'For me personally I believe that humanising our relationships is one of the most important ways that we can build a better world and yet we have this technology and some of the best and brightest brains in the world working on technology that does exactly the opposite.'

There was no question then, for him, about needing to do something about Pine Gap, but 'what to do' was 'a much harder question to answer'. He believed in 'the power of social movements to create change', had studied their history; he cited the examples of anti-nuclear protesters who stopped nuclear ships coming to Australia, who stopped the export of uranium for nuclear weapons, and of people resisting the Vietnam War.

'Putting all these things together' was the basis for his belief that the Pilgrims' action at Pine Gap would be 'something worthwhile', even if it seemed 'small'. He used the word 'symbolic', which I was sure McHugh would jump on, and spoke of his friends as 'ordinary people resisting war' with what they had: 'It's not about sitting at home and doing something, it's about going to where it happens and trying to disrupt it ... We had to break through the euphemisms and the obfuscation of a place like Pine Gap and we had to say "People die from this place – we're coming here to lament and we're coming here to resist it." And that's what we did.'

McHugh's first questions, as with Margaret, went to Andy's participation in lawful protest and pushes for change. McHugh mentioned his interest in the conscription referendums of 1916 and '17: 'I'm quite familiar with the story, I can elaborate more if you would like,' offered Andy. McHugh declined; his interest was to make a point about the referendums as an example of the democratic process.

Andy made his own point: 'It was also a process of extraordinary repression … a law was passed that stopped people from circulating any anti-conscription material … And many people resisted and went to gaol, including a future prime minister. John Curtin actually went to gaol as an anti conscription protester.'

In any event, said McHugh, it was 'an example of people protesting and change occurring', overlooking that part of the protest had involved confrontation with the law of the day.

He moved on, sure enough, to the Pilgrims' 'symbolic action of lament'.

'A lament is a recognition that death occurs there,' said Andy. 'It wasn't purely symbolic.'

But he had written about it as 'a symbolic action', put McHugh.

Andy said that the intent to disrupt was 'implicit' in what he had written, given that it was for the Close Pine Gap website.

McHugh's next set of questions went to the real purpose of the action, including the live-streaming; it was all about bringing people's attention to the issues; the idea of disruption was 'at best, a vain hope'.

Andy spoke about the intended disruption of his attempted lock-on to the bus earlier that day: 'We failed that but, you know, you try and stop a war. You can't let one little defeat get in the way and so we went back again the next night … we have limited abilities, but it was a real attempt to disrupt it in whatever way that we could.'

'That wouldn't have stopped the signals coming through, would it?' put McHugh. 'You knew it was in vain, isn't that right?'

'Do you think I would be here, risking the years in prison that I am, if I thought this was vain?' asked Andy. 'I had a real belief in the power of that action.'

If all the Pilgrims likely impressed the jury with their principled determination, I expect it was Franz, next into the witness box, who most stirred them. The effect came not from any evocation of the horrors of war – the material the Crown was so concerned to keep from the jury – but from his own simplicity, commitment and compassion.

He spoke of how he had been raised on a Catholic Worker farm, how he lived now in a house of hospitality 'trying to make the world a better place for those in need'; of his belief in the power that music has in evoking human emotion, in spiritual worship and prayer, in self-expression and 'ultimately trying to create a change'.

He and his co-accused friends 'were drawn together out of a very mutual concern for what we knew Pine Gap to be'. He and Margaret had channelled their 'musical passion' to compose the lament 'in a way that paid tribute to those innocent lives that had been destroyed or heavily impacted on by Pine Gap's operations'.

He described the walk into the base on that beautiful cold night, reaching the 'cattle fence that we knew to be the perimeter fence of this stolen Indigenous land that the US claims as their own'; their surprise that police patrols were waiting for them on the other side; their determination to keep going in the process they believed in, 'that involves lamenting and mourning and considering the change that needs to come'.

With 'the power of the spirit' behind them they crossed the last fence 'one way or another' and started up Mariner's Hill. He described Margaret's final surge of energy, them both playing as they went, and finally reaching a rock 'overlooking this base that was right in front of us – an amazing sight. This kind of monstrosity laid out before us, lit in bright red lighting, and we just played and it was a very emotional moment, you know. We had this thing in our sights and all we could do was just play.'

His voice trembled.

'It felt like quite a long time, I'm not sure how long, but eventually we were pulled off and arrested … I don't know what else to say about that. But we were, I was at least, drawn to take part in this act out of a very strong desire that I have to stand up for the people that are least deserving of the destruction and misery that is caused to them.'

Franz wept then.

McHugh would have been relieved that the lunch break was due and he didn't have to start in straight away on taking apart this evidence.

After lunch, Franz had a bit more to say, about his awareness of Pine Gap and its activities through his father as well as his own research, and in particular about the impact of the documentary *Drone*, by Tonje Hessen Schei, the one that had screened ahead of their action and had been seen by Federal Agent Davey among others. He sought to play an excerpt for the court. McHugh objected along the usual lines. Reeves allowed it but decided against its tender. Still, the jury had seen it and Franz got to speak to it, how he and his friends could not fathom any other way to oppose the strikes than by going to the place 'on our soil where we allow these acts to happen': 'You saw earlier Scott Ludlam come in and he spent however many years in parliament always talking about Pine Gap, like, for as long as he had been in this position of power, and nothing happened.'

McHugh was fairly gentle, and the questions were now well-rehearsed, until he got to challenging Franz on knowing there were 'other reasonable ways to respond'.

'Honestly,' Franz said, 'if the police had not reached us after crossing that first perimeter fence, nothing would have stopped me from that point from scaling the next fence and trying to disrupt the base and I see that as being very reasonable for the crimes that were being committed.'

'That's your belief in that regard, but that was never part of the plan though, was it?'

'—We never discussed it, no,' said Franz.

Tim followed Franz onto the stand. He was less emotional but struck a similar note of soft-spoken gentleness, kindness, sincerity.

His Catholic parents had introduced him from a young age to teachings of Jesus about nonviolence and love of your neighbour, 'which, considering the state of the world sometimes, is quite a radical sort of teaching'.

He had become aware of the Catholic Worker movement, which was all about 'living simply so you increase the chances of other people being able to just simply live'. Through this movement he was also exposed to political activism, and in New Zealand, where he was from, the example of the Waihopai Three (including his cousin Sam Land) loomed large.

They had sought to 'expose a similar installation like the one here at Pine Gap': after a lot of 'praying and meditating', they deflated one of the balloons that cover these 'ginormous satellite dishes that gather information in a very similar way to Pine Gap, yes, to spy on the world for the US government basically'. A jury acquitted them, 'so that's a happy story of how political activism is successful in some ways'. That held lessons for him: 'It has made me aware firstly, that we were allowed to protest ... It also made me aware that you can take this argument to the court and win, quite obviously, from that example.'

His awareness of 'the particular injustice happening here at Pine Gap' came later when he met people in Australia, including his current housemates. He 'began to sense the urgency of the situation', exposed to similar information as Andy and Franz, including that coming from whistleblowers, a term that he liked: 'I've never actually delved into its origins but I assume it means when someone blows a whistle on a sports field, you're pointing out something that's happened, something has gone wrong – stop.' The wrong, of course, was killing people which goes against the most 'intimate ... part of our being'.

All this went to the frame of mind and beliefs that led to him being at Pine Gap that night: 'I come from a rural area in New Zealand ... and I just kept thinking, you know, there's the people who are being affected most by this ... just simple farming people like myself. And I just was thinking of the farm or the farmer, on that sort of level, the very sort of primal need to just intervene in any way possible sort of overtook me, I suppose, and that was all the reason that I needed to take action.'

McHugh questioned Tim on the Waihopai action, hoping to draw the distinction between it and what the Pilgrims had done. It was not part of the plan to go on down and attack a radome at Pine Gap, was it?

'I hadn't thought of doing it but I can't rule out the possibility that it would've, might've happened if we'd gotten any closer.'

Jim had yet to take the stand, but the Pilgrims asked if Richard Tanter could give his evidence as it was now well into the afternoon and he too had to fly out the next day.

McHugh attempted to object, reiterating the Crown's concern 'about opinion evidence, hearsay, speculation, conjecture and all the rest of it', but Reeves remained consistent: he couldn't rule on the evidence until he heard it.

One of Andy's first questions to Richard was about Andy interviewing him for his radio show in 2016. Richard summarised what they had talked about: Pine Gap's critical role in US planning for certain aspects of nuclear war; its expanded functions over the last ten to fifteen years in US non-nuclear military operations; and the implications of this for Australia, one being whether or not Pine Gap is 'a very high target in the event of major war' between the US and, for example, either Russia or China; another being for the integration of Pine Gap into US military planning, which raised 'questions of sovereignty' for Australia.

Andy wanted to tender a transcript of the interview as evidence. McHugh objected. He was sure Professor Tanter was well qualified to speak 'on some subject matters', but the Crown could not deal with this evidence.

Andy argued that it went to the Pilgrims' state of mind, and to 'the objective reasonableness' of their beliefs 'in terms of defence of others', given the imminent threat that Pine Gap poses to them.

Reeves was stern: 'I told you this morning that Australia's foreign policy, its alliance with America, its defence policy, are not in issue in this proceeding and I will not allow evidence directed to those broad subjects. It seems to me that that's exactly what the Professor has just talked about.'

Andy withdrew the attempted tender: 'You can keep your copy there, Mr McHugh. It's good reading.'

Reeves went on: 'Those matters are determined by the parliament and by the executive, by the government. Courts don't deal with those matters.'

Andy asked Richard what the basis was of his knowledge about Pine Gap. He answered with an account of his relationship with the late Des Ball and with Bill Robinson, and the technical and historical papers they had published: 'There's nobody else that's done that kind of work.'

It was 'the classic bootstrap', objected McHugh, the knowledge was based 'on the witness's own researches'.

'It's qualifying the expert, isn't it, Mr McHugh?' asked Reeves.

'Yes, but what can the Crown do with it, your Honour? We can't cross-examine in the dark.'

Reeves asked Andy what the evidence would be directed to.

His peer-reviewed research 'gives a bit more credibility' to the Pilgrims' beliefs and speaks to 'the imminence of the threat, that it was a real threat'.

'None of these questions has yet been directed to any of those things,' said Reeves.

So Andy asked, and Richard started to answer, about Pine Gap's interception systems. Objection: the witness's answer could only be based on expert evidence when beliefs were not in dispute, it would be 'prejudicial for the Crown' and 'likely to be emotive'.

Reeves allowed the evidence, but said it should be 'confined and direct'.

Richard spoke about Pine Gap's space-based and ground-based interception of communications across a wide part of the world, which make it the most important platform for that activity. Downlinked to Pine Gap, a 'very large part' of the signals are processed and analysed there, and the data on 'targets of interest' are passed on to the National Security Agency and from there into the wider US military and presidential targeting arrangements for drones.

'Would the US drone bombing program be able to function without Pine Gap?' asked Andy.

'Certainly not at the level it does at the moment.'

Objection: 'This evidence still needs to be tied in to the knowledge of the accused.'

Reeves again told Andy to confine the evidence to the defences. Andy repeated his question and Richard took the answer a little further: 'The effects that we are seeing, which I certainly know you and I talked about, in countries with which we are not at war – Pakistan, Somalia, Yemen for example – would not be nearly so great.'

'And so if someone were to disrupt the functioning of Pine Gap, would that have an effect on the US drone program?'

It would interrupt 'to a very limited extent ... that targeting escalator of information'. The other effect would be on public opinion: 'Polls in Australia have shown that not only yourselves but many Australians are very concerned about the illegality of these matters and I suspect, had there been disruption of a major scale, it would have been very effective in drawing attention to that.'

'Has there been an escalation in the drone program and the use of Pine Gap?'

Yes. The number of US drone strikes had escalated and the documents revealed by Edward Snowdon 'conclusively confirmed what we had inferred from other sources, that Pine Gap is an essential part of that program and it has been commended for its role in that program'.

Andy then attempted to take Richard to questions of Pine Gap's role in nuclear targeting and as a possible target itself. That was never going to get very far.

'Your Honour,' began McHugh.

'Yes, I know,' said Reeves, repeating his comments about issues of broader public policy not being the interest, or prerogative, of the court.

Andy withdrew. Jim, more willing to be argumentative, wanted to ask a question but Reeves wouldn't allow it. There was no cross-examination: Richard could leave.

Taking the stand, Jim talked a little about his life as a Catholic Worker and his commitment to nonviolence, before trying to show a clip from a 1987 film, Bob Plasto's *Journey to an American Spy Base*. In it, a former CIA agent talks quite frankly about 'how the Australian people were lied to' over what the base does. Reeves wouldn't allow it – too long ago, too remote from the defences they had outlined.

'Now could we get to your evidence about your beliefs and how they affected what you did last year?' he asked.

'Well, that certainly goes to the evidence of my beliefs, that six minute film, but if you don't want to see it, that's fine. Not much I can do about that,' Jim answered.

He didn't get any further with a 2003 article by Robert Fisk, 'the famous war journalist, writing about civilians killed by cluster bombs', the targeting of which Jim thought 'Pine Gap was quite possibly involved in'.

'Can we get to the point?' Reeves was getting impatient. 'Your evidence is directed to the incident last year. Let's hear about that.'

After a break, Jim focussed his attention on more recent events. He had been made more aware of drone strikes by the American activist Kathy Kelly, who had stayed in their Brisbane Catholic Worker house for a while. She had lived in Iraq in the war zone, she had visited victims of drone strikes in Pakistan and Afghanistan and been involved in nonviolent civil disobedience in the United States, protesting the drone program. Jim had read Richard Tanter's articles and exchanged emails with him, so he was well aware of Pine Gap's role in drone strikes. He wanted to tender an article by Eric Margolis, describing the role of the CIA and the Pentagon in inflaming the situation in Syria, dated 3 September 2016.

'Did you read it then?' asked Reeves.

Jim wasn't sure; if he hadn't read that article, he had read similar. Against McHugh's objection, Reeves allowed it as 'an article of the kind' Jim had been reading and going to his 'state of mind' in 2016.

Jim turned then – no doubt to Reeves's relief – to the action at Pine Gap. He spoke of seeking permission to go onto the land from its Arrernte custodian, or caretaker, Peter Coco Wallace. 'Naturally, when I was asked if I had a permit, in my record of interview that you saw, I said yes, I did. I had a verbal permit from a person I believe is the true owner of that land.' There had never been any agreement to take it, or buy it, or to sign a treaty or anything, so he couldn't see how 'the Commonwealth or the US government can be considered rightful owners of that land in any way whatsoever. So as far as I'm concerned I had a permit to be there.'

He spoke about the poster he had carried on to the base, showing the dead child in Basra: 'Pine Gap might have been involved in that child's destruction and if not, it was certainly involved in many other children's destruction.' He wanted to tender it, despite the Crown's previous objections.

On what issue would it be tendered? asked Reeves.

'It tells the true nature of Pine Gap's role in killing, it shows the nature of what war is about, what Pine Gap is about and we carried it on that night ... to expose Pine Gap.'

'The accused has said himself he doesn't know that it's associated with Pine Gap,' objected McHugh, adding that it was 'only meant to inflame the emotions of this jury'.

Reeves felt that objection could be dealt with by direction and allowed the tender. At last, a small win for Jim.

McHugh went first for what he saw as a weakness in Jim's point about the lack of a treaty signing Pine Gap over to its current users.

'Are you aware of any such treaty ... or document of title from Arrernte peoples for this building?'

'No, no.'

'Or for land in Alice Springs?'

'No, I didn't—'

'There is, in fact,' Reeves cut in, referring to the native title determination over Alice Springs.

McHugh took it in his stride. The *Native Title Act* is an Australian law, was Jim familiar with it and its dealings with pastoral leases? And was he aware the area surrounding Pine Gap is a pastoral lease?

On the outside, yes, but he didn't know about 'inside Pine Gap itself'.

'You're not sure that that area was once a pastoral lease and what its legal status was?' ventured McHugh (with his own rather incomplete picture of the land title history of the base).

Jim wouldn't be drawn; as far as he was concerned it belonged to the Eastern Arrernte.

~

After the trial I spoke about these issues with Peter Coco Wallace Peltharre, the same senior Arrernte man Jim had approached for permission. I was surprised to learn from him that there had been some consultation with his grandfather and uncle about where to build the base (which contradicts what Bryan Law understood he had been told by Patrick Hayes, and what Jim understood to be the case). I asked first though how he felt about the base being on his traditional country.

'It's okay,' he said, 'they're not doing any damage for my grandfather's sacred site [Kweyernpe] … They're doing a good job, they're looking after the site as well while we're not there. We just get there sometimes, not too often, but they told us, it's okay they can look after it as well, in case of any damage.'

Are there sites inside the fences he and his family want to visit?

'No, no, no, they [the base] kept away from that. My grandfather was here, my uncle was here at that time, when they started the military base there. They told [the authorities] that "You can't do it here, you've got to be a bit further that way."'

This was right at the beginning?

'Yeah, right at the beginning, I was only young.'

So it was before Land Rights?

'All before that.'

I asked too what he thought about the issues raised by the protesters, about what goes on inside Pine Gap, not in relation to sacred sites, but what the military base does.

'I don't know nothing about what's happening there either,' he said. 'It's okay. I'm a kwertengerle, a caretaker for that sacred area, it's my responsibility. If I want to go there, have a look around, I can go there. But I don't know what's going on inside the facility at Pine Gap. I've never been there.'

I understood, I put to him, that some protesters (not only Jim) in the past had come to see him to ask his permission to go inside the base, to protest about what it does.

'I said what's the good about protests,' he replied. 'Well they tried to stop all that. I said I don't think we can stop it, might as well leave it as it is, for safety's sake, you know, they look after us, from the space base over there, they look after us.'

Did he mean the base looks after all Australians?

'Yes, exactly. In case, you know, something from the other side of the world they might send, but they look around, they keep watch all the time ... That's why we're safe here ... I told them, well they are trying to save our lives, they are trying to look after the country ... Why protest? I wasn't quite sure about that.'

One of things the protesters are worried about, I told him, is that information collected at the base is given to the American military and it is used to kill people in other countries.

'Oh I see,' said Peltharre, 'I didn't know about that, I didn't know about that. They must have hurt someone. They never told me much. Oh, they have killed people from the other side of the world from here?'

'Yes.'

'Well, why is that?'

'That's the question,' I said. 'That's what the protesters are asking, why are we doing that, should we be doing that?'

On a later occasion we discussed all this again. I showed him, by way of illustration, surveillance images of before and after a drone strike in North Waziristan, published in *The Washington Post*. He sees 'a lot of this on the news', he said, and asked, 'Do they read their minds in there, see who they are, if they are innocent people?' And 'I think the protesters have to go on, to try to stop all that. Otherwise we never know what will happen here next ... If they want to do it from America, they can do it from there, from their own country.'

~

McHugh asked Jim if he took the same view about Alice Springs as he did about Pine Gap, that it's owned by the Arrernte people.

'I guess so, if there's no written agreement to hand it over, yes, sure.'

So, there were some laws that he acknowledged and others he didn't?

'Well, Martin Luther King Junior said, "An unjust law is no law at all". I tend to agree with that.'

Who decides whether the laws are just or unjust?

'Well, that's what we're given a conscience for, isn't it? I mean, the Nuremberg trials made a good ruling on that, I think.'

McHugh turned to the action itself.

Jim said they went to the base to lament and resist war crimes, with nonviolence being the only way 'open to Christians' of doing that.

So at 4 am on 29 September last year, how was that nonviolent action going to close down Pine Gap?

'We entered the base in an attempt to disrupt the base, as had been done in 2005,' said Jim, '... and although Mr Napier claims that there was no disruption to the killing business as usual, we were not aware of that at the time ...'

In 2005 he cut the fence, but in 2016 he didn't take any fence-cutting material with him, did he?

No, agreed Jim.

He would have known that the police would stop them very quickly, as they had done a few days earlier.

No, said Jim, and with some reason: Paul Christie had been inside the base for five hours, and in 2005 he and Adele had managed – 'miracles happen' – to get inside the compound itself: 'It was equally possible we could have gone all the way to the fences, climbed over the fences' as he'd seen an elderly nun do way back in '87 and a man, the one Margaret had mentioned, in 2006.

Either McHugh was getting tired or Jim was managing to side-step his questions; he didn't manage to get Jim on a pin like he had Margaret. The questions started to go all over the place and Jim started to enjoy himself. McHugh, trying to get him to acknowledge lawful avenues for political action, asked whether he had ever considered taking the government to court 'on any of these sorts of matters'? He wasn't expecting Jim's answer: 'I did do a citizen's arrest once of Peter Dutton over the war crimes issue, in 2004 I think.'

'How did that end up, I have to ask – how did that end up?'

'Well, he was embarrassed but other than that he wasn't punished for his war crimes at all, unfortunately.'

That had not advanced McHugh's attack very far but he soldiered on: 'In any event you know there are legal avenues that may be potentially available to you if … for instance if you have accusations of war crimes?'

'I've never known a court yet to overrule the laws appointed by the government to protect their military and wars … People have tried to take George Bush, John Howard to the International Court but they haven't got very far, and I myself don't have the money, expertise or whatever to do that, so other than that I'm not sure what we could do legally.'

'Alright, but in terms of political power, I think you heard Ms Pestorius say she was in the Greens. Are you involved in any political parties?

'No, no, I don't vote myself.'

Jim had mentioned the court system, was he aware of the separation of powers?

'Yes, yes, it's a bit of a mess in some ways, isn't it?'

'Well, you say that. In any event, you understand the government does not run the judiciary?'

Jim came back: When he and his friends were prosecuted for going into Pine Gap in 2005, they won on appeal on the basis that 'the prosecution had to prove that Pine Gap was necessary for the defence of Australia and so our conviction was overturned. But then the government immediately, a few months later, changed the law. So the separation of powers is only a very temporary thing in an instance like that, then the government will just change the law to suit themselves, so where is the separation of powers … in that instance?'

'Well I won't give a lecture on … in fact I withdraw that.' But then McHugh tried to extract some advantage from the governments's amendments to the *DSU Act*, getting nowhere. His junior, Dobraszczyk, must have muttered as much.

'Just excuse me, Your Honour …' He bent towards her, then straightened, 'That's the cross-examination, Your Honour.'

~

The jury was sent home and the legal argument began on whether the Pilgrims had met the 'evidentiary burden' of their defences, that is, were they 'reasonably possible'? The Crown would say no. For their action to be in defence of another, 'they'd have to think that if they shut the base down for a minute or hours that that means that there'll be a stopping of signals which would result in the stopping of bombing in Iraq or Syria or Yemen or these places, these names that have been bandied about'.

As for the circumstances of emergency, 'these things are ongoing and, if they're ongoing ... it's ordinary, it's not sudden or extraordinary. Callous as that may sound, Your Honour, into the ears of the accused, but on their own evidence this is happening all the time.'

Even if the emergency existed, the defence would also fail in proving the action was the only reasonable way to deal with it. There were other reasonable ways to respond, all those lawful actions that McHugh had said were 'a hallmark of our democracy'. And as well the accused knew the conduct could not be in defence of another because it was a lament and because they knew they would be arrested – 'it wasn't for disruption'.

It was the end of a long day. All of that would be traversed again in the morning and the Pilgrims would have their opportunity to reply.

CONSCIENCE VERSUS REASON

22 November 2017

The jury came in late morning but were told to go away again until at least after lunch – legal argument was still being heard. Reeves then asked McHugh to clarify the Crown's concession on the beliefs of the accused – all those times when he had said the Crown accepted that the beliefs were genuinely held.

That concession was made, McHugh explained, in order to avoid evidence about those beliefs being given, but of course it had been given anyway.

Reeves said he would treat it accordingly and not for any admissions as to truth or accuracy. The ruling in any case would not turn on the content of the beliefs, but rather on an objective assessment of them – an exercise he characterised as 'murky'.

McHugh spoke about the necessity defence and the 'rare circumstances' in which it is made out. 'The classic example is a father with a child having a snake bite, and running a red light to hospital.' In that situation there is 'a close temporal connection' between the emergency and the conduct. In the Pilgrims' case, it was 'inappropriate', he argued, to try and fit the facts into a defence of necessity: 'They just do not work and they certainly don't work for public policy reasons ...' The circumstances could not be seen as overwhelmingly impelling disobedience to the law. He cited former High Court Chief Justice Gleeson on the limits of the defence – historically it had required circumstances of 'urgency and immediacy'.

'You jumped over the part that I thought was most directly relevant to this case,' Reeves said. 'That is, "The law cannot leave people free to choose for themselves which laws they will obey or to construct and apply their own set of values inconsistent with those implicit in the law."'

That was clearly going to be his overriding guiding principle.

Under the necessity defence, the Pilgrims would also have had to believe that their action was the only reasonable way to respond. There were other ways, McHugh argued, in line with his cross-examinations, 'legal avenues' that the Pilgrims had taken and were continuing to take – 'protests, gaining media exposure and so on'. So they could not believe 'that the only way was to enter the prohibited lands on that night'. The lament was 'an extension of the protest they'd been performing that week' and they assumed they would be arrested: 'It meant that their actions were in fact the actions of martyrs. These were voluntary decisions ...'

He cited Lord Hoffmann, 'a wonderful speech' on civil disobedience. While Paul Christie had invoked it as an argument for leniency in sentencing, McHugh was using it, and the British cases it was examining, to highlight the 'voluntariness' of the Pilgrims' conduct.

Reeves pointed out a difference with the British cases, that the Pilgrims had done no damage, but McHugh raised in turn the Pilgrims' evidence that they intended to cause disruption, implying intended damage of a sort. Once again he was having it both ways, a 'vain hope' on the one hand, serious intention on the other.

He went on: 'If they're out there simply lamenting and playing their musical instruments, and that's what their design was, in order to be arrested, to be martyrs, to continue the process, to come to this court, the Supreme Court, rather than having it dealt with in a magistracy ... we say that's not an objectively reasonable response to the circumstances.'

There was overreach in that argument: 'You can't take into account what's happened since the conduct,' said Reeves.

McHugh returned to the eloquence of Lord Hoffman on unlawful conduct in the context of 'a functioning state in which legal disputes can be peacefully submitted to the courts, and disputes over what should be law or government policy can be submitted to the arbiter of the democratic process'. In such circumstances, a jury should not be able to consider acts of civil disobedience as justified.

This line of thought would also be adopted by Reeves.

Since the day before and now well into the morning, the exchange had been entirely between McHugh and Reeves. Margaret, busily taking notes, was beginning to wonder if she would be heard.

When Reeves was ready for her, she did her best, speaking on behalf of them all, to persuade him to allow their evidence to go to the jury: some of the issues raised by the prosecution went to 'ordinary definitions and ordinary reasonableness' and the jury should be the ones to assess that, she argued. For instance, that an extraordinary emergency existed: its 'longevity' and 'the sustained attacks on community' were 'part of its extraordinary nature'; its existence was 'far more than speculative', with 'a lot of evidence on the public record that the harms are occurring'. This included Richard Tanter's evidence. He is 'the most highly regarded academic on the subject', she said. 'I mean, it's a pity that we couldn't get somebody from inside the base to come and speak to what they're doing, but they don't tend to do that.'

She tried to counter McHugh's logic, that the ongoing nature of the strikes made them ordinary. In the context of an escalating overall emergency, each individual event was also an emergency, she argued: 'Just as if you have an emergency department, you know, if the same person comes in every night with domestic violence, you don't go, oh well, you've already been in twenty times this month, and so it's not an emergency, is it?' So if, as in Afghanistan at the time, there were 'three strikes a day and increasing, that is an imminent peril'.

She wasn't sure of being on track. 'Am I doing the right thing?' she asked Reeves.

Yes, he said, but she wasn't yet dealing with their 'biggest problem', which was that committing the offence had to be the *only* reasonable way to deal with the emergency.

She pointed to there having been no change in policy despite hundreds, thousands of people, including themselves, attempting to get some sort of change by other means: 'It's looking like the only reasonable way is to go in there and switch a button, or do whatever you need to do to shut that thing down, you know, to stop the signals being sent.'

Of course, they had not managed to do that. If they had done, their defence might have had legs.

Margaret made a point about the way Australians are taken into wars (thinking no doubt about former prime minister John Howard taking the country into war in Iraq): 'You know, you have to remember that we live in a country in which there is a non-reviewable exercise of executive or prerogative power to wage war. Those decisions can be made by one, or two, or three people in the government, without regard to the parliament and without even regard to their own caucus ... no discussion, nothing. And we've seen that happen.'

In this political context, the Pilgrims' action was 'reasonable conduct', proportionate to the seriousness of the emergency, 'way worse than the way we're trying to address it', and proportionate to their ability as citizens. 'You know, we do what we can do' but it has to go beyond 'doing what's not working, over and over'.

On the issue of taking the law into their own hands, she argued that this was 'the whole idea' of the legal defences, that sometimes such action was justified: 'You know, it was said over and over at Nuremberg that you don't just sit there and watch things happen ... And the people in Pine Gap, we wish, were attending to the Nuremberg principles. But, they are not. So, we have to. There's an emergency that requires us to go in and do something about that.'

'Nuremberg, Ms Pestorius,' Reeves interrupted, 'was directed to a totalitarian dictatorship, not a functioning liberal democracy.'

'Yes, Your Honour. And?'

'Well, those principles were applied to address that situation.'

Reeves was right about the liberal democracies not being on trial in Nuremberg. The aerial bombing of whole cities, in which those democracies had excelled, was not an issue examined in the Nuremberg trials nor in those held in Tokyo. After the decimation of German cities by Allied bombs (far more extensive than the Blitz on English cities), after the fire-bombing of Tokyo and other Japanese cities, and finally the nuclear bombing of Hiroshima and Nagasaki, examination of the German and Japanese air wars would have been intolerably uncomfortable for the Allies. Victors' justice compromised from the start the 'universality' of the Nuremberg principles, but they are still an articulation of an ideal, however imperfectly applied. This is what Margaret was appealing to.

'As I understand,' she said to Reeves, 'they addressed war crimes and the right of the ordinary person to address war crimes. And, we're saying war crimes are occurring here and ordinary people need to rise up ...' She noted Lord Hoffmann's comments, from the same famous opinion, that in times of war 'citizens are entitled, indeed required to refuse to participate in war crimes'. She then undermined her argument somewhat, adding 'And, are we in war? Are we not in war? I don't know.'

In terms of acting in defence of others, Margaret spoke first of their subjective belief: 'We're called. We go for different reasons. We're at different points of life ... I think we've been able to give a good overview of that.' But there was also an objective basis to those beliefs, she said: firsthand accounts of people arriving back from Afghanistan and from Iraq, victims telling their stories, whistleblowers, veterans who had joined the peace movement, soldiers 'who don't believe in war': 'So it was clear that the harms were occurring.'

Moving from belief to their action and its intent to disrupt, she said: 'You know, we have young men who can climb. They can climb anything and, you know, the age and gender and state of health which goes to this, of these circumstances, need to be left to a jury. A jury will understand that, you know.'

As an action it was 'very different' from the protest by David Burgess, a case McHugh had referred to. With fellow activist Will Saunders, Burgess was responsible for the 'No war' graffiti on the Sydney Opera House just before the invasion of Iraq: 'We have a place that's directly doing the harm,' said Margaret, 'not like Burgess who was on an opera house. We're saving lives. We're stopping violence. We're stopping extreme harm.'

If McHugh overreached at times, so did Margaret here on the difference between what they had broadly hoped to do and what they actually managed or plausibly had the means to do. The young men might have been able to scale the fence and even scale a building in the compound, but to actually disrupt operations, nonviolently, would have been quite a leap.

McHugh had raised Burgess because the trial judge in that case had removed self-defence (and defence of others) from the jury's consideration; there was an appeal on that issue but it failed on the basis that the conduct he and Saunders were claiming to defend themselves from was lawful. This was another critical weak point for the Pilgrims' defences – the presumed lawfulness of Pine Gap and its activities. Margaret argued that 'even somebody like Malcolm Fraser was questioning the lawfulness' as were 'a number of human rights lawyers around that time'. She suggested that the prosecution should have to prove Pine Gap's lawfulness.

On the contrary, Reeves said, the Crown in relation to the 'defence of others' was relying on the subsection that excluded the defence if 'the person is responding to lawful conduct': 'Let me put it to you this way, whatever is happening at Pine Gap is happening because of, in part, decisions of an Australian Government to allow it to happen. Those decisions are prima facie lawful.'

The Pilgrims were saying, Margaret countered, that the presumed lawfulness was contested and if it was so contested, it needed to go to the jury: 'Apart from this sort of general blanket of prima facie "we excuse everything that goes on there", the Australians have not showed any legal opinion about that.' The Pilgrims didn't know what Pine Gap was doing was lawful: 'We believed it was unlawful. And we had knowledge to that effect, from a range of eminent sources,' she said, citing Fraser again and his belief that drone warfare would eventually be recognised as unlawful activity.

~

In his 'Reasons for Ruling', published later, Reeves characterised the Pilgrims' burden of evidence as 'reasonably lenient'. Nonetheless, it had to cover a lot of ground. If they failed on any of the points of the necessity defence, they failed on the whole defence.

The ruling dispensed with what he had seen at first as 'a murky area' – assessing 'objectively' the reasonableness of the accused's beliefs, a difficult undertaking especially in a contentious matter of public policy where value judgments are involved. Judges in analogous cases had shown the way, by usually assuming that the accused's beliefs were reasonably held. Reeves took the same approach, accepting that the Pilgrims reasonably believed that an extraordinary emergency existed, but questions remained for the other points.

On these, he found that the Pilgrims had failed, 'even on the view most favourable' to them. It was clear that the trespass was not the only way to respond to the perceived emergency. Their evidence showed that both before and after, they had actively pursued 'lawful protest activities'. None of them had identified any critical matter that meant the situation was different on the date of the trespass. They chose their course of action that day because they deemed their other activities 'ineffective'. The purpose of the trespass was 'to gain greater public attention' for their views. It was not 'in any real sense' a response to the extraordinary emergency; they were not 'in any real sense' impelled to break the law.

His reasoning was much the same in considering whether they had acted in defence of others (and protection of their property). The test for belief in the need to act was subjective, but there needed to be 'a reasonable possibility that their response to the threat was reasonable in the circumstances as they perceived them to be'. Again Reeves found that there was no evidence to support that.

He did not mention the issue of public policy and the presumption of Pine Gap's lawfulness.

~

The headline I gave to my report for the *Alice Springs News* that night jumped to a conclusion: 'No extraordinary emergency at Pine Gap: judge rules'. Like everyone else, I did not yet have the benefit of Reeves's published reasons. With hindsight, I could have more accurately headlined, 'Judge accepts belief that extraordinary emergency existed at Pine Gap'. I would have had to qualify that in the article, that it was only for the purposes of the ruling, but there would have been an interesting point to ponder – the acknowledgment of the value judgments involved and the contentiousness of Pine Gap and its activities, which put the question beyond legal reasoning.

When the jury returned after the legal argument had concluded, Reeves simply told them that he had ruled against the legal defences the Pilgrims had sought to make out. This meant the Crown did not have to disprove the defences because there was 'no reasonable basis' for the jury to consider them.

All of the parties, in their closing addresses and summing up, addressed the broad issue of civil disobedience, with McHugh and Reeves both emphasising its limits and reminding the jury that they had to decide the verdict with their heads, not hearts, while the Pilgrims appealed to them to follow their consciences.

Reeves was more potent on the issues than McHugh, as was to be expected perhaps given that he was the arbiter of questions of law. McHugh muddied the waters by bringing in 'what the suffragettes did in giving women the vote in the early 1900s. Australia was one of the first countries in the world, I think, to allow that. There were civil disobedience matters in respect of those occurrences. Of course, the law has changed and so it should be.' That sounded, if anything, like an argument *for* civil disobedience, rather than one to impress upon the jury its limits.

Reeves's comments would not be heard till the next day; the five Pilgrims came first. He emphasised to them that in final submissions they should not address any of the matters relating to the legal defences: 'If you do that, I will try to stop you and if I don't stop you, I will certainly be directing the jury to ignore it, but don't take that as an invitation to do it.'

All that was left for the jury to decide were the facts proving the charges. Andy tried to use that to his rhetorical advantage, although he was no doubt doing exactly what Reeves had told them all not to do:

> I appreciate that the court thinks it's an important process that everyday people assess the facts in front of them … We also believe that this is a very important process. And I guess it's part of why we're here. Because we are people who became aware of certain facts and decided that there were things that we needed to do in response … I think every person, in order to do what's right and to live a life that's free and that you can be proud of, needs to be able to see the facts and judge what to do about them.

Franz appealed to them similarly: 'Our fate is now in your hands and your hands alone. You know what we did, you know why we did it …' He wished they had been able to do more, to prevent the deaths by drone strike in Shadal Bazar – that was his 'only regret'.

Tim asked them to use their collective life experience 'to decide whether or not it was a crime to nonviolently endeavour to shut down a facility that unleashes untold horrors on innocent people on a daily basis'.

Jim urged them to follow their consciences: 'Nobody can make you find us guilty or not guilty … If you think we're guilty of a crime here, then sure, find us guilty. But if you think the real crime lays elsewhere, I'd ask you to find us not guilty.'

Margaret was last to her feet. Most of the time she attended court in her habitual shorts and t-shirts. Now she had put on her wedding dress, to draw into the court a sense of Bryan, their marriage and their work for a better world. The dress was printed on bodice, sleeves and hem with birds, fish and suns – a wearable hymn to creation.

She spoke of the love and support the Pilgrims had found in the Catholic community in Alice Springs. She quoted Martin Luther King: 'It's no longer a choice between violence and nonviolence, but between nonviolence and nonexistence.' She spoke of the cycles of 'grief, silence, denial, war' that she encounters in her work as a therapist: 'My clients go in and out of gaol … If I must join them, then so be it. There's much work to be done in the gaols of Alice Springs, I've heard.'

She spoke of Pine Gap as a prohibited area for 'ordinary people': 'It's a law defined by successive governments, with no vision for peace. We don't, we can't challenge that. We haven't challenged that. We had that taken away from us, much earlier than this court case. We can only stand in our conscience, in our sense of what is right and we also ask you ordinary people to stand strong in your conscience. Your sense of what is right as you reflect on grief and war and silence and denial. There's a crack in everything – that's how the light gets in.'

There she was quoting Leonard Cohen; now she quoted Bryan, from his closing address in the 2007 trial:

> And I am here because we were one. What's important for me in your deliberations is that you follow your conscience and do the right thing. I'll abide by that result. I'm happy to pay the price of my actions and I call upon you and I call upon everyone else in this country to stand up for what you believe is right. If we do that, he said 'my father', but I say my husband would be proud if he was still alive. I'll be proud. Our son will be proud and we'll build the kind of community that I think this country aspires to.

23 November 2019

The jury had all night to be troubled by the Pilgrims' appeals to both their consciences and emotions. Margaret in particular had been very moving.

Emotional reactions were quite natural, Reeves told them, but they had to put them aside. They were legally required to decide only whether the Crown had proven beyond reasonable doubt the facts relating to the charges:

> Just as it is not open to the accused to choose whether to comply with the laws of this country, it is not open to you, consistent with your duties as jurors in this case, to find them not guilty, not because the Crown has failed to prove all the elements of the

> charge against them, but because you think the law under which they have been charged is unfair to them or unreasonable or too harsh.
>
> You will probably be left with no doubt about the sincerity of the beliefs held by the accused and the strength with which they hold them. Nevertheless, those beliefs, however sincerely and strongly held, do not override their obligations as citizens of this country to comply with the laws made by our democratically elected parliaments. This is one of the fundamental tenets of the operation of the rule of law under which we all live.

Where his analysis of the elements of the crime involved questions of law, his directions were binding on them: they had to apply the law as he outlined it to them. This included a direction that 'the area of the Joint Defence Facility Pine Gap was, as at 29 September 2016, a prohibited area for the purposes of the Defence (Special Undertakings) Act'. They were obliged to accept that this element of the charge was established beyond reasonable doubt. (This was the effect of the 2009 amendments to the law.)

He defined what recklessness meant in relation to the knowledge or awareness of Pine Gap being a prohibited area. Awareness or knowledge of the fact were entirely subjective but being reckless as to that knowledge – that is, being aware of a substantial risk that Pine Gap was a prohibited area – involved an objective factor for the jury to assess: was it justifiable for the person to take that risk?

Regarding the evidence, they had to exclude, Reeves told them, all of the oral testimony of Scott Ludlam and Richard Tanter, and most of the evidence-in-chief of the Pilgrims as well as several of their exhibits. This must have been rather crushing for the Pilgrims to hear, even if expected.

He broke down the questions the jury needed to answer on the evidence:

- Did each of the five accused enter the prohibited area?
- Did they each intend to enter the prohibited area?
- Did they each know the area was a prohibited area? Or, were they each aware there was a risk the area was prohibited and

the risk was substantial, and having that awareness, were they each unjustified in taking that risk?
- Additionally, for Andy, did he possess a photographic apparatus in the prohibited area, and did he intend to?

He commented that none of the accused 'expressed any regret for what they had done'. I wondered whether that was relevant to fact finding, but perhaps it could be drawn upon for inference about intent.

On Jim's appeal that if they thought the real crime lay elsewhere, they should return a verdict of not guilty, Reeves acknowledged again the conflicted feelings they might be experiencing. He thought comments by the former Chief Justice of Australia would help them understand why they should put any sympathies for the Pilgrims aside. When he made the comment, Murray Gleeson was chief justice of the New South Wales Supreme Court and dealing with a case not dissimilar to this one. During legal argument Reeves and McHugh had had an exchange over the first part, about people not being free to choose for themselves which laws they will obey, but not its corollary for the jury: 'Nor can the law encourage juries to exercise a power to dispense with compliance with the law where they consider disobedience to be reasonable on the ground that the conduct of an accused person serves some value higher than that implicit in the law which is disobeyed.'

The jury retired to consider their verdict. It was not quite midday. The case against each defendant had to be dealt with separately so they were always going to take longer than the half-hour Paul's jury had given him. Everyone went away to get some lunch but soon we were all back again – the Pilgrims, McHugh and co., the AFP officers, the journalists – waiting in the lobby for the judge's associate to call us in. There's always a sense of drama at this time in a trial. It's just like the movies show it, a nervous energy infecting everyone, a degree of relief that most of the arduous business of the trial is over, mixed to a greater degree with apprehension about its upshot.

The Pilgrims were talking and laughing together, a little more keyed up than usual. I hadn't spoken much with any of them at that stage, except for asking the odd clarifying question, but I had noted more than once how much they seemed to like each other's company, how upbeat they mostly were, joyful even. Margaret especially seemed to foster this, keeping their morale up, always joking about, even though she probably took the trial more seriously than the others, put more into it, with Andy a close second.

I made the most of the wait by writing up my report for that night; all I would need to add was the verdict, if it came. I had a headline in reserve: 'Even if "unfair, unreasonable or too harsh", it is still the law' and in the end this is what I used. At 4 pm the judge's associate called us in. Margaret rushed to the ladies, to put on her wedding dress again, but she needn't have worried. The jury wasn't ready and didn't think they would be before 4.30; they asked to be allowed to return the next day. They also asked to hear again parts of Andy's and Tim's cross-examinations. They were certainly being conscientious. With Andy, the evidence was to do with what he had said about filming and live-streaming: he had been quite frank about it and had agreed that the intention was to go as far as they could onto the base and that he expected the police to come. With Tim, he had agreed that they had entered the base and that they expected to be arrested. If the outstanding questions for the jury were around intent and awareness of the risk of the base being a prohibited area, any doubts were now probably settled.

24 November 2017

The jury resumed deliberations at 10.30 and by 11 they sent word they were ready. I sat behind the Pilgrims. Franz had taken a Bible from his backpack and allowed it to fall open. His eye went to this passage in Luke and he read it out to his friends:

> He said to the crowd: 'When you see a cloud rising in the west, immediately you say, "It's going to rain," and it does. And when the south wind blows, you say, "It's going to be hot," and it is. Hypocrites! You know how to interpret the appearance of the earth and the sky. How is it that you don't know how to interpret this present time?

> 'Why don't you judge for yourselves what is right? As you are going with your adversary to the magistrate, try hard to be reconciled on the way, or your adversary may drag you off to the judge, and the judge turn you over to the officer, and the officer throw you into prison. I tell you, you will not get out until you have paid the last penny.'

They were all delighted with how apposite it seemed, especially these questions from Jesus, 'How is it that you don't know how to interpret this present time? Why don't you judge for yourselves what is right?' – the same questions that they were asking of everybody with regard to Pine Gap. I was more focussed on what followed: the spectre of judgment and prison.

I didn't envy the jury foreperson her task: six times she had to say guilty, six times to agree that that was the verdict of them all. It was like the tolling of a mournful bell.

After I posted news of the verdict, 'Jack' went back to my report of the day before, about Reeves's directions to the jury. He started his comment by quoting from it: 'It is not open to them, as a matter of law, to find the accused not guilty because they believed the law under which they are being judged is "unfair, unreasonable or too harsh". Well actually,' he wrote, 'this option is open to the jury and if I had been on the jury I would have taken it. In the USA there are many not guilty findings irrespective of the law and the evidence where the three strike law jailing offenders for life applies. In the years to come we as a society may wish we had paid a lot more attention to the cause of the peace activists.'

I talked to Russell Goldflam about this later. He noted the legal parlance for referring to such verdicts, that they are seen as either 'perverse' or 'merciful', depending on the point of view, but they cannot be overturned. He was unaware, however, of any civil disobedience cases in Australia, unlike in New Zealand, the UK and the US, in which a jury has returned a merciful verdict.

~

There was nothing merciful in McHugh's submissions on sentencing. This was expected, of course, after he had called for the mild-mannered Paul to be sent to gaol. The five had put up much more of a fight in court.

He left no stone unturned to make the case for the offending to be seen as serious. Although only Andy had been charged and found guilty of the 'photographic apparatus' offence, McHugh proposed that the 'live-streaming' was part of the 'overall circumstances' of the offending for all five.

That might have been so, said Reeves, if the film had shown some feature of the base, but they had only filmed themselves, an 'entirely different' matter.

They had potentially struck 'at the heart of national security', put McHugh.

How so? asked Reeves.

The context of the legislation made it so, argued McHugh, with its maximum penalty of seven years for the offence.

There was no evidence beyond reasonable doubt for that level of seriousness, countered Reeves; someone 'blowing up the base' might attract the maximum, as a 'worst case'.

The legislation allotted the penalty to 'mere entry', argued McHugh. That showed how seriously the offence is viewed in contrast to an ordinary trespass offence with a maximum of two years.

Andy's additional offence required 'some sort of accumulation' in sentencing, he proposed, again making a point about the legislation: the offending wasn't on 'just any land'; the 'rights and wrongs' of that were not for him, McHugh, to go into.

Jim's long history of offences of a 'similar character' was a relevant consideration, McHugh argued.

Why were they seen as 'similar'? asked Reeves. There might be a comparison with his Rockhampton conviction (given it was on a defence base), but why would he treat the wilful damage to cemeteries as a similar offence? (This conviction had been particularised in the Crown submissions for all four men.)

McHugh pointed to what that offending was about – the removal of the sword from the cross – and that it had occurred after the

Pine Gap offending, after the attorney-general had consented to the prosecution, and after summonses had been issued. This meant that Reeves should take into account specific deterrence, including for Franz, even though the Pine Gap trespass was his first offence. Margaret too had been involved in sustained offending over a long period of time, so specific deterrence should play a role in her sentence.

McHugh acknowledged the Pilgrims' strong Christian beliefs and ideals. The source of their beliefs didn't matter for sentencing considerations, but it was relevant that they were 'strongly held', especially in the case of Margaret and Jim, because of the pattern of their offending going back years. Jim, in particular, 'was on the street and will continue to offend' whereas the others had 'paused from time to time'.

The Crown wanted custodial sentences imposed, he said, with a portion of the term actually served. In terms of general deterrence (the effect of the sentence on people other than the offenders), a custodial sentence's 'full bite' was only felt when time was actually served.

McHugh commented on Andy and Tim as 'intelligent' and 'relatively young' committed nonviolent activists. Despite their previous convictions, their prospects for rehabilitation were better than those of Margaret and Jim, and in sentencing Reeves should have regard to directing their energies and abilities away from criminal activities.

Franz too was a committed nonviolent activist but was likely to have been influenced by his father, which could be an element in mitigation. 'As he matures he may come to other views,' suggested McHugh. (Franz would tell me later how cross that made him: 'I'm my own person,' he said.)

McHugh didn't want to say much about the offenders' references, other than to turn against them the ones attesting to the deep Christian beliefs that motivated their actions. This did 'not bode well' for their prospects of rehabilitation: 'There is a real risk that Margaret Pestorius and James Dowling will continue to break laws they disagree with.'

Neither did he want to say much about the submissions Margaret had made on behalf of them all, but he proceeded to make quite a few points anyway. On the argument that their offending was less serious than the 2005 trespass, Reeves would need to make a finding on intention. McHugh had said earlier that they intended to disrupt the base – as was often the case, trying to have it both ways – but Reeves had responded that this was not proven beyond reasonable doubt.

On the 'delay' in their prosecution, which the Pilgrims said had been left hanging over their heads for some months, it was minimal and should be disregarded. On the use of the *DSU Act* as opposed to ordinary trespass legislation, the submission should be rejected as the accused did not consent to having the matter dealt with summarily – they had wanted a jury to hear them. On their 'conscientious motives' being taken into account, along the lines argued for by Lord Hoffman, that might apply to first-time offenders, but there comes a time when such leniency is no longer applicable; that time was 'well past' with respect to Margaret and Jim.

Andy responded on behalf of them all. It was 'quite surprising' to hear that their offending struck 'at the heart of national security' when the prosecution had spent that past week arguing against precisely that (any such idea was supposedly fanciful). Theirs was an 'act of civil disobedience' and the police knew that they were 'not a significant threat'. (This too was an argument having it both ways.)

Andy suggested that the Crown's logic on specific deterrence for those with a long history of civil disobedience was faulty. Civil disobedience would not be deterred by penalty; it would be deterred by laws that were 'not unjust'.

The Crown had claimed there were 'no comparative cases of offences under the DSU Act' and that analogous offences such as trespass on Commonwealth land were not helpful because of the much more lenient penalties. Andy argued that the penalties (fines) handed down by Justice Thomas to the Christians Against All Terrorism were a guide, as were the fines given to the hundreds of trespassers at Pine Gap over the decades.

On the Crown's submission that imprisonment with a non-parole period was the only appropriate penalty in the circumstances, Andy said that there was a wide range of sentencing options available and that a term of imprisonment should be used only as a last resort.

McHugh interjected that there are 'different forms of imprisonment'. He had already noted that home detention was not available in Queensland and had argued that actual time be served, but he had also noted that they had each spent a day in custody. His written submission, however, had discussed non-parole periods – such a period would be required were the sentence to exceed three years.

Andy then spoke of himself: his previous convictions were all for 'principled' actions; he didn't break the law 'for any old reason'. The Pine Gap action was 'done out of conscience', out of 'deep respect' for all humankind, and was without malicious intent towards anybody.

Reeves enquired about his personal and financial circumstances, and about contrition. He lived 'minimally', Andy replied, believing that 'money is the root of all evil'. As for contrition: 'I don't have much to offer.'

Reeves made some comments about sentencing principles before Margaret began. Past sentences only became relevant if they disclosed a pattern, he said. There was also a difference between sentencing young, first-time offenders and people with long lists of prior offences, who, being older, more mature, 'should have more sense'.

'Only one police officer thought I was not young,' joked Margaret, 'and he was wrong – but it was a dark night.' Most people laughed. Then she made a serious point, quoting the appeal judges in the Christians Against All Terrorism case, that their trespass was at the lower end of offending. It had no victims, it did no harm, 'we are not dangerous in any sense of the word'.

The relative seriousness of the offending is one of the sentencing principles to take into account, replied Reeves.

She contested the suggestion that she had a sustained criminal history. Apart from a few traffic matters, they were all civil disobedience offences, and for these there had been a long hiatus:

between 1996 and 2006, being a responsible parent, she hadn't had time to get arrested.

Reeves said only analogous offences would be taken into account.

She challenged the Crown submission that a term of imprisonment was warranted because she was likely to continue to break laws that were against her beliefs. It made her 'very concerned for dissent in this court', that that should be considered a 'normal response' when she was 'wholly nonviolent'.

Reeves said McHugh had pointed out that imprisonment could include, for example, a suspended sentence: 'You seem to be assuming it means being locked up.'

'He asked for real gaol time to be served,' replied Margaret, with reason. She argued that this was an 'unworthy restriction on democracy', that democracy doesn't just get developed in the legislature. People breaking unjust laws – around slavery and genocide, for example – have pushed their countries towards just laws. 'People stood up.'

Her personal circumstances included providing daily care for someone dying from cancer; going away would be problematic, as there was no-one else really available to do that. She had a good part-time job, but had been on unpaid carer's leave for the past year.

Reeves asked her about contrition, and also whether the mobile phone confiscated during the trespass was hers and what should be done with it.

She opposed forfeiture of the phone, for which the Crown had applied. If it was destroyed that would be waste, she said, it was 'not an ice pipe', it could be a useful phone for somebody. She had 'nothing to say at all about contrition, except perhaps before God, for not doing more'.

Franz told Reeves he had a job as a part-time support worker for a man who required twenty-four hour care but he would not call himself poor. He was in his third year of a four-year double degree, 'whatever that counts for', and he had community commitments, including to a street kitchen on Friday nights and other things to try 'to make the world a better place'.

He closed by quoting Dorothy Day, co-founder of the Catholic Worker movement, 'the greatest woman who ever lived': 'All of our problems stem from our acceptance of this filthy, rotten system.' She was a 'strong character', said Franz, with a little chuckle. He added that he could not be 'a slave to the law of man as long as those laws are able to be broken at whim by those in power, resulting in the mass killing of innocents'.

Reeves took this last statement as a direction on contrition.

Tim, like Franz and Andy, had been a support worker, but was presently unemployed and planning to return to live in New Zealand, where his circumstances would be much the same, he said. His previous offences were all principled nonviolent actions responding to 'the insatiable blood lust of the Commonwealth'.

Finally, it was Jim's turn. He had acted 'in good conscience to resist war': 'Anyone with their eyes open could see that Pine Gap was engaged in serious war crimes' and, according to the universally relevant Nuremberg principles, 'people acting to obey the law is no excuse when it comes to war crimes'. In arguments hardly designed to win over Reeves, he referred to the writings of Ingo Müller on the complicity of the judiciary in Nazi Germany, making a comparison with the NT Supreme Court, that instead of punishing those committing war crimes, it was considering how to punish those resisting them.

Reeves didn't bite then but he would bring the point up in his sentencing remarks. He asked about Jim's financial circumstances and contrition. Jim said simply that he didn't have much money, and as for contrition, he was 'extremely sorry Pine Gap was still involved in serious war crimes'.

Sentencing of all six Peace Pilgrims, including Paul, was set down for 4 December in Brisbane, where Reeves normally resides. The Brisbane four would have to front the Federal Court there; Margaret and Paul could appear by video-link from Cairns. Tim, who had optimistically booked a ticket back to New Zealand for 30 November, would have to cancel it: he was obliged to appear in an Australian court for sentencing.

There was a bit of discussion between Reeves and McHugh about jurisdiction. Bail arrangements could be extended to allow the Pilgrims to travel back to Queensland. If sentences of full-time custody were imposed, they would be taken into custody there before being brought back to the Northern Territory, where the sentences would initially be served until transfer requests were processed. Someone would appear at sentencing for the Crown, but it wouldn't be McHugh.

~

The number of Pilgrims supporters had dwindled. Graeme Dunstan and those travelling with him in the Peacebus had gone, so there was no longer a colourful presence outside the court; and there had never been many locals following the trial, whether as supporters or simply interested. Scholar–advocate Felicity Ruby had left, so had the other journalists with the exception of the ABC. The mood was deflated. It was over but not quite, and now there was the long journey home to get underway. I asked to get a photo of all six Pilgrims together and the ABC wanted an interview.

We straggled downstairs into the laneway that runs in front of the court. Someone went to get a banner and the instruments. Parsons Street, at the end of the lane, was unusually busy, with a steady flow of people heading towards the mall. It was the tail end of the White Ribbon march, I realised, an annual event in the campaign to end men's violence towards women. Senior police had joined Aboriginal activists at the head of the march and hundreds followed them, including other police officers, welfare professionals and many ordinary citizens. It was comforting to realise that the lack of local interest in the verdict for the Pilgrims was not solely down to complacency. A co-signed letter to the editor of the *Alice Springs News* later made the connection, thanking the marchers and the Peace Pilgrims for 'breaking the silence violence'. They wrote:

> With amazing coincidence, the White Ribbon Day march passed in front of our new Alice Springs Supreme Court at the very same

> time as the five Peace Pilgrims were found guilty of trespassing on land used by the Joint Defence Facility Pine Gap. The jury had been instructed to not consider the defence argument used by the Peace Pilgrims, that they had acted to protect many innocent people who continue to be at risk of the escalating emergency created by drone warfare in many countries around our world.

A banner materialised. It was one that had done long service, designed by Bryan Law. 'Pine Gap Terror Base' it proclaimed, 'Civilians Bombed While U Wait, "back into the stone-age".' That phrase recalled an infamous suggestion by American four-star general Curtis LeMay for the US bombing campaign in North Vietnam. He was by then well practised in the art, having led the incendiary bombing of Tokyo and been the commander who relayed President Truman's order to drop nuclear bombs on Hiroshima and Nagasaki.

Franz and Margaret tuned their instruments and played the lament for the dead of war. Franz was so right about the power of music to provoke emotion: heads sank as its sadness enveloped them. For once, no-one was laughing or joking. They didn't want to stay there though, especially not if it could look like an air of defeat in the photo. They bucked themselves up with a few verses of 'Down by the Riverside', Franz tapping his foot and Margaret swinging into her bowing. Then they sang 'The Vine and the Fig Tree', the low chant, 'And every one 'neath their vine and fig tree / Shall live in peace and unafraid', and the swelling chorus, 'And into plowshares turn their swords / Nations shall learn war no more'. It was poignant, full of the hope that keeps them going. It might have been then that the Pilgrims took up residence in a corner of my mind.

FIELDS OF ACTION

As the Pilgrims travelled north, I travelled south – my annual pilgrimage to see family. I was in Adelaide on 4 December when they were sentenced, watching out for posts from #ClosePineGap to learn of their fate. Around lunchtime the news came through: Reeves had rejected the Crown's call to send them to gaol, imposing fines instead, $1250 for Franz and Tim, $2000 for Paul, $2500 for Andy, $3500 for Margaret, $5000 for Jim, convictions recorded for everyone. The courtroom in Brisbane had erupted in applause and cheers from their supporters.

Photos were soon posted. Margaret and Paul stood with a small group of supporters, among the palm trees outside the Cairns courthouse. Margaret was wearing her wedding dress and they carried placards – 'No choice on war for aUStralia', 'Pine Gap hardwired into US war crimes'. Both of them were looking at the camera, smiling, not triumphantly but, it seemed to me, with unwavering purpose. In Brisbane, a larger crowd gathered outside the Federal Court, with colourful peace banners and placards, the four men standing together in front. Andy was smiling broadly, hands thrust into his pockets, his t-shirt reading, 'Rise up against war – Pine Gap on trial 2017'. Jim was reaching up to put his arm around Franz's shoulders – Franz is a good head taller. Jim's t-shirt read 'No war, Life is sacred'. He couldn't have looked happier alongside his son, whose strong arms were folded and head high, a smile playing on his lips. Only Tim had a more inward look, not unhappy at all, just his usual greater reserve.

Reeves had been sanguine on penalty relative to the Crown's call for imprisonment and actual time to be served. Under the *Crimes Act*, for imprisonment to be imposed it had to be the only appropriate sentence, and clearly he thought it was not. The fines were steeper than those imposed on the Christians Against All Terrorism; indeed his minimum for the Pilgrims had been the Christians' maximum.

He had leaned towards gaol for Jim, given the extent of his previous offending – twenty-seven convictions for trespassing on Commonwealth land or similar offences, going back to 1986. McHugh had argued for sentences that would signal to 'would-be protesters that trespassing on facilities such as these crosses the boundary between legitimate protest and criminal activity and that the potential publicity to their cause from such activities is neutralised by the risk of severe punishment'. The word 'neutralised' made for a strange argument, given the likelihood that severe punishment for such mild acts of civil disobedience would have provoked outrage. Reeves saw this: 'If I imprison you,' he said to Jim, 'I think that will be likely to make you a martyr to your cause rather than to underscore the law-breaking in which you were involved. I think you and others will use it in that way among that section of the public who hold views similar to yours about the use to which Pine Gap is put.'

He also noted that there had not been a significant increase in trespass offences at Pine Gap, further weakening McHugh's argument. Indeed the most recent previous instance that he had been informed of had been the 2005 case. He didn't say so, but it seemed a deterrent effect had been achieved not so much by penalty imposed but by prosecution under the *DSU Act* (with its threat of severe penalties), much as Justice Thomas had observed in 2007.

He made some stern remarks, as I thought he would, about Jim's submission that the court should not punish those who are resisting war crimes: 'I say in respect of that submission, Mr Dowling, that the declaration of Pine Gap as a prohibited area is not even remotely a war crime. Your offending did not therefore even remotely concern a war crime.' This seemed to misconstrue what Jim had been saying was a war crime, but Reeves's ruling against the defences had stripped away all that evidence from the case; all that was left, in the legal sense, were the facts around trespass – it was all he needed to refer to.

He was also stern towards both Jim and Margaret as 'the older members of the group'. Addressing Margaret, he said: 'You should be setting a good example as law-abiding citizens, in particular to younger protesters who are attracted to your cause, but instead, both you and Mr Dowling seem to be determined to set examples as trespassing, law-breaking protesters.'

Speaking to Paul, although his comments could surely apply to all of them, he recognised that he had shown 'a sense of proportion' and had not caused 'excessive damage or inconvenience'. He referred to his citation of Lord Hoffman and the restraint shown by the courts in conscientious civil disobedience cases. Reeves's own sentencing appeared to reflect his assent to this approach.

On their claims of the effectiveness of civil disobedience in changing unjust laws, he suggested 'that may be so of some instances in the more distant past'. This overlooked at least one prominent example playing out at the time. Health professionals speaking out about conditions for asylum seekers in Manus Island and Nauru were doing so in breach of the *Australian Border Force Act 2015*, maximum penalty two years' imprisonment. Apart from a campaign of civil disobedience – they continued to speak out, some daring the government to prosecute (nobody was) – Doctors for Refugees launched a High Court challenge 'on the grounds of unjustifiable burden on the freedom of political communication'. Following this, amendments were enacted to relax the Act's secrecy provisions in relation to health professionals. This happened in 2017, in the months before the Pilgrims' trial. (In early 2019, a fairer law for medical evacuations of people in offshore immigration detention was also enacted, although the Coalition government managed to repeal it in December 2019.)

Other examples of recent effective civil disobedience are legion, unless Reeves regards his own younger adult years as 'the more distant past'. Some 1400 people were arrested and gaoled in the Franklin River blockade of the early 1980s, to name one of Australia's most famous civil disobedience campaigns, ultimately successful in preventing the damming of the river. It led to the birth of a political party, the Greens, and contributed to bringing the Hawke Labor government to power in 1983. Hawke wasted no time in passing legislation to protect the river as a World Heritage Site and successfully pursued the Act's enforcement against the state of Tasmania, backed by the High Court. The draft resistance movement during the Vietnam years was another example of a mass-participation civil disobedience campaign that ended in achieving its goal. It became one of the issues that brought the Whitlam Labor government to power in 1972. Whitlam immediately abolished the draft by administrative action, and it was terminated by legislation the following year.

So there was little reason for the Pilgrims to agree with Reeves's general argument around lawful protest, but as nonviolent direct action practitioners, they were of course ready to accept the consequences of their principled law-breaking.

Reeves had prefaced all his comments with a general one about the *DSU Act*, that its maximum penalties had remained unchanged 'during the sixty-five years of its existence', despite numerous other amendments. This indicated that the Australian Parliament regarded their offences 'as particularly serious ones' even if they were at 'the lowest end of the scale of breaches' under that Act. They were 'an unlawful but relatively harmless extension' of their 'peaceful and lawful protest activities'. None of them had any implements or any other means of 'causing any harm to the facility itself, let alone Australia's national security'.

He had also stressed in advance that he was not sentencing them for exercising their right 'to publicly express the grave concerns' they had about Pine Gap's activities as they believed them to be. He was sentencing them because they deliberately chose to break the law. The message the sentence had to give to the broader public (general deterrence) was that protesters should exercise their free speech rights lawfully.

He referred to the Crown's submission that 'protection of the public' was a relevant sentencing principle in the case, but he had difficulty in seeing that: 'Indeed, on the evidence the persons whose safety was most directly under threat in these particular trespasses were the five of you wandering around in very rough and hilly terrain at night time.'

Their lack of contrition was relevant, but he noted that their offences were 'devoid' of victims and, apart from the expense caused to the public purse, they had not caused any damage. He also thought they may have misunderstood what they were expected to be contrite about: they were not being asked to express contrition for holding their views, but rather for 'making the deliberate choice to break the law in order to give expression to those views'. Even if he had explained that in court, I doubt very much that any of the Pilgrims would have replied differently on contrition.

He rejected the Crown's submission about them electing to have a jury trial rather than being dealt with in the lower court: that was their right and he didn't see any reason why they should be penalised for exercising it.

Margaret and Paul had indicated that they would be able to pay their fines; the four Catholic Workers were unlikely to. But not long after sentencing, they were all told to bin the paperwork they had signed; it would be reissued. At the time of writing, it had not been. It seems that the issue of jurisdiction, which had been discussed between Reeves and McHugh, has been the stumbling block: they had been tried under Commonwealth legislation in the NT Supreme Court; sentencing took place on the premises of the Brisbane Federal Court; it was all too convoluted.

On the day, though, the Pilgrims gave the penalties little thought. They were glad that the draconian calls of the Crown had been rejected and that the case was behind them, but the central achievement of the whole ordeal for them was that they had 'dragged Pine Gap into the light'.

~

The late Des Ball recognised the national interest in some of Pine Gap's activities, as does Richard Tanter. If Richard were in an intelligence organisation, he'd be wanting to use Pine Gap's capacities to watch, for example, some extremist Islamist organisations in Indonesia and the Philippines. Keeping an eye on them 'is a legitimate use of that function of Pine Gap'. Ball also called for this in his last public interview, asking why part of Pine Gap's capacity was not directed full-time to trying to track members of Jemaah Islamiyah. Australia's processes, however, needed to be governed by 'rules, principles and procedures', he argued. 'Capture, arrests, warrants, evidence. We should leave the killings to the CIA and JSOC cowboys at Menwith Hill.'

Is hosting Pine Gap an all-or-nothing situation? I asked Richard. Would it be possible to pick and choose, for Australia to say we want Pine Gap to do this but not that?

Theoretically, yes, Richard said. The answer would lie in the tasking schedules, given that Australia has equal representation at Pine Gap, working in every part of the facility except for the American cryptography room (and the Americans don't work in ours):

> The Australian Government could say we do not want the tasking schedules of the intelligence satellites or infra-red satellites to be doing X and Y and we want a verifiable way of doing that. We've got our people in the room, at that meeting – we *have* a verifiable way of doing that. And on things like providing data to assist targeting in countries in which neither Australia nor the US is at war, in other words blatantly illegal extrajudicial executions, I don't see why that's not a perfectly reasonable objective for Australian governments to aim for.

But Australian governments show no sign of being worried about that, do they? I'm thinking of Scott Ludlam's tussles with former attorney-general Brandis, of my own non-answers from former defence minister Pyne.

'At ministerial level, no,' said Richard. 'I suspect there are some people in Defence who do, who would think these things may be against the Australian national interest properly considered, which would include respect for international law. What it means is most easily discovered by thinking about what we are going to say when the Chinese government starts doing this. They are going to start using drones to knock off Uighurs and people over the border in Kazakhstan and places like that. I think we are not going to like that.'

There is another area of Pine Gap's activities that Richard believes would be possible to close down, and which would be essential if Australia wanted to become compliant with the Treaty on the Prohibition of Nuclear Weapons. The treaty opened for signature at United Nations headquarters in New York on 20 September 2017. At the time of writing it had been signed by eighty-one countries and ratified by thirty-six of them; it needs ratification by fifty before coming into force.

Major political change would be necessary for Australia to embrace the treaty's aims. Christopher Pyne, for example, was quite explicit about Australia's commitment to 'the US policy of extended nuclear deterrence': the nation is not only a beneficiary but 'an active supporter' of the policy, he said in his statement on the joint facilities. If a shift eventuated and compliance with the ban treaty became the goal, then the relay ground station in Pine Gap's western compound would need to go. Its provision of early warning of a nuclear attack is also essential for US nuclear retaliation. The same technology that detects the heat blooms of an adversary's missile launches also allows the US to compile targeting data for a retaliatory strike – on the missile sites that haven't fired yet. This function makes the relay ground station incompatible with the treaty's prohibition on assistance in the use or threat of use of nuclear weapons. Richard believes it could be closed without compromise to what the US holds is in its national security interest, because there is already technological redundancy in the system – with satellite-to-satellite crosslinks, Pine Gap can be bypassed.

~

The Pilgrims went back to their lives and may not return to Pine Gap. There's plenty of peacework to do closer to home.

Jim and Anne will soon have to leave the Peter Maurin Farm, with their lease set to expire. They are unfazed. All their children will be independent by then, and they're confident 'God has a plan for them whatever that may be'.

I visited them at the farm after the trial. The house, an old Queenslander on stilts, is small – three bedrooms for nine people when all their children were at home – but it's functional and has its charms, like the back deck Jim built out into the treetops where I found him reading Dostoyevsky's *Possessed* (borrowed from the library) over his morning coffee. The coffee beans were past their use-by-date, salvaged from the coffee roaster where his third-born son Ben works, but the coffee tasted fine. The vegetable garden, fruit orchard and poultry provide the foundation of the family's meals, otherwise supplemented mostly by what Jim harvests from supermarket dumpsters. He has ingeniously developed the farm's large degree of energy self-sufficiency, including a biogas toilet that fuels their stove and a solar cooker perched on a pole so that it is just outside the kitchen door – on a fine day it can bake a loaf of bread on sun-power alone. Solar hot water, without a booster from mains power, means no hot showers on cloudy or wet days, but fortunately this is subtropical Queensland.

His way of living demands hard work – the tending of the gardens, the dumpster-diving, the endless upkeep of their systems, including the vehicles with their biofuel adaptations. On the upside, it has freed Jim largely from wage labour; he can choose how he spends his time. Just one twenty-four hour shift a week brings in more than the small amount of cash he needs and he gives the rest away. For decades he also pursued a cottage business, making soap, but recently handed it over to one of the residents of Dorothy Day House. In term time Anne, with her own children mostly grown, drives the local school bus and does some relief teaching; she also works part-time as a creative arts therapist. Between them there is enough money, even for what they call luxuries: a rare movie, an occasional holiday, and for Anne, sometimes a coffee with friends.

'People look for luxuries if they are not satisfied with day-to-day life,' Jim told me. 'I can remember one time sitting in the lounge room with two of the young children climbing on me and laughing, and feeling happy and thinking, life doesn't get much better than this really! I live in a house surrounded by trees and sunshine, or rain, I have plenty of food, don't work too hard, have a beautiful family. Life is a luxury – when I look at it positively.'

Franz was watering a new garden bed along the side of Dorothy Day House when I met him there one morning. Having been raised on the farm, he knows about growing food. It had taken some time to build up the soil. The seedlings were mostly vegetables, but there were some flowers too, rescued from a dumpster.

Inside the floors were freshly mopped and everything was tidy: there had been a rental inspection the day before. Andy was downstairs at the computer, writing a music review, typing it up from a handwritten draft in a notebook. I sat with Franz upstairs at the large kitchen table, a vase of flowers at one end in front of open casement windows. Benches on either side could each seat four, six at a squeeze, and quite often there are a dozen or more people for dinner.

While we were talking, his older sister Rebekah arrived. Franz's mood had been low lately, and she'd come to take him up the coast for a couple of days' reprieve. She'd founded the house and lived there for a number of years, so she knows well the strains of it, including having little private space. Anne, who had been visiting her mother in the city, arrived shortly afterwards. She hugged Franz tightly. She was keen to see him get away.

I asked her later how she feels about the risk of Franz going to gaol if he keeps up his nonviolent civil disobedience.

'Of course I don't want him to go to gaol and I'll be praying for him every day. But I think he'll be fine, he's a strong man, he's a kind man, prisoners will like him. I've been in gaol, we're in Australia, it's not too bad, though it's never not a risk. In my experience, if you're kind to other prisoners, listen to their stories, you make some good friends. And sometimes the hand of God is there.'

In 2019 it wasn't gaol but a punitive Queensland Government bureaucracy that would put Franz to the test. As part of the final requirements for his degree, he had lined up a Creative Industries internship, leading guitar workshops at a community centre. Because it involved working with young people, he needed to renew his Blue Card. The renewal was rejected: his Pine Gap trespass conviction and his lack of regret for his actions were seen as raising concerns 'regarding his ability to exercise proper restraint and self-control, judge appropriate behaviour and present as an appropriate role model to children and young people'. It is worth remembering that Justice Reeves took the Pilgrims' lack of contrition into account in sentencing, as he was bound to. It is a relevant factor in relation to the personal deterrence effect of a sentence. In Franz's case he determined that a fine of $1250 reflected 'the required specific and general deterrence effects'. Now here was Franz, getting on in exemplary fashion with study and work and his many community activities – some of which, like leading his church choir, also require a Blue Card – facing an onerous extrajudicial layer of penalty.

At the time of writing, Franz was appealing the decision in the Queensland Civil and Administrative Tribunal, with the pro bono assistance of two lawyers. Ironically, his chances of success would be greater in the Northern Territory where it is unlawful to discriminate against a person on the ground of an irrelevant criminal record. This is not the case in Queensland. Still, Franz was hopeful for a positive resolution of his case and that it would put in place fairer guidelines for dealing with nonviolent civil disobedience convictions. (With Extinction Rebellion activism gaining momentum, there promise to be many.) While waiting for his appeal to finalise, he was pushing away opportunities for arrestable actions and sought out a very different experience for finishing his degree: a study tour in Chennai, India. Collaborating with Indian musicians, he immersed himself in South Indian culture and landscape and in 'understanding how some of the world's poorest people are living'.

I caught up with Andy again, preparing for the Food Not Bombs street kitchen. Part of an international movement, Food Not Bombs grew out of anti-nuclear activism in the US in the 1980s. Its political aims have broadened into a commitment to nonviolent direct action for social change in whatever contexts its self-determining chapters face, but it's always associated with providing free vegetarian and vegan food. In the US and some other countries its volunteers have encountered significant repression, even persecution, but the movement has flourished. Today it has over a thousand chapters around the world. It's easy to see its fit for Andy and his Catholic Worker housemates. In many ways they seem to operate like a 24/7 Food Not Bombs.

The place where they prepare the food – a house that is used as a community space by not-for-profit groups – is a couple of blocks from the busy main street of West End. A dozen or so people were sitting round a big table, talking, laughing and chopping. Andy was sorting through a small mountain of tomatoes, throwing the rotten ones into a compost bin, thinking about what he'd make with the rest. He introduced me to the people around the table then promptly disappeared.

Soon I was making headway on my own small mountain, apples for a crumble that Jess had peeled while she kept up a lively banter with everyone. They were all very aware of the Pine Gap action and the trial. Jess said more than once how wrong it all was; she'd written letters to politicians at the time. She lives three days a week at Dorothy Day House, the rest of the time with her mother. She finds company at the house with people her age, and a sense of purpose. She spoke freely of her application to the National Disability Insurance Scheme, of a recent visit to hospital, her medication, her anxiety. A lot of that talk was addressed to Holly, who was preparing a gigantic mixed salad, and her brother, who sat back from the table, mostly quiet and not involved with the food preparation. He too was applying for the NDIS, a little further along in the process than Jess. Holly was studying naturopathy and planning to travel overseas at the end of the year. She used to live in Dorothy Day House, but now she was sharing a flat with a friend. 'Are you a Catholic Worker?' I asked, trying to figure out the dynamics around the table. 'I consider myself one,' she said. Jess asked her about being a Christian versus being a Catholic – does it make a difference to how close you are to God? No, said Holly.

One of the Dowlings arrived, Ben, the son who worked for the coffee roaster. He knew everyone, laughed and joked around with them all before starting in on some chopping. His short dreads were pulled up in a hair-tie. Jess called him 'pineapple head', delighted with herself.

Andy returned. He'd used some money from the kitty to buy soy sauce and garlic from the Chinese grocer. There was talk about a charity footy game between musicians and media that he was going to take part in. He loves footy, question was, would he play with the musos or the media?

A newcomer, who had seen their notice on Facebook, was helping Andy chop tomatoes and garlic. Often the approach to the cooking is to throw in whatever ingredients are available and hope for the best, but Andy wanted to cook the tomatoes without too much else in the mix, so you could taste them. Michael came out from the kitchen. He is one of the six core Dorothy Day House residents, having turned away from a career in IT. Andy was starting a bit late, he said, this always happened. They wanted to be set up by 6.30, he reminded everyone.

Suddenly at 6.10 they were ready. Even the washing-up was done. They loaded the food in its large serving dishes onto a trestle table with the legs folded up, along with plates and cutlery. Eight of them picked it up, four either side. Andy told them they had plenty of time, not to hurry, and so began the march to the main street in West End. I walked behind, watching the tilt of the table, fearing that at any moment the dishes would slide off, but they were practised at this. It was a strange and moving procession, past the restaurants and bars getting ready for the evening trade. People looked up as they went by, but no-one seemed surprised; it happens every Friday.

There was a 'Free Food' banner on the wall in a small open area near the public toilets, rather grandly called People's Park. Some regulars, some newcomers were already waiting, a mixed bunch, not necessarily homeless, if you can judge that by appearance, but stretched in different ways, including socially. Company is part of what is on offer and quite often after the meal there'll be music, especially when Franz is there. The table was set up in no time and people began to help themselves, including everyone who had prepared the food. Most people sat on the ground to eat; some stayed silent over their plates; others started to talk. Andy was warm and easy with the newcomers. I sensed the advantage of his bare feet and worn clothes, felt the barrier of my jacket, boots and shoulder-bag of stuff.

In the 2018–19 summer Andy left Dorothy Day House, heading to Western Australia, slowly. He came through Alice Springs in the January and, with friends, climbed up to the top of the range from where you can see Pine Gap in the distance. It's a place full of memories for him now.

Western Australia, when he eventually got there, didn't hold him. There was no Catholic Worker community for him to land in, but in any case he was looking for something different to do. Next I heard he was back on the east coast, sailing towards Manus Island with a group intent on protesting Australia's asylum seeker policies. He wasn't sure though how far he would go with this venture. Bad weather had them stalled in Cairns and for Andy, looking around for ways to direct his activist energy, the Adani protest camp seemed like a pretty good alternative. Camp Binbee, as it's known, became his temporary home. Not without its frustrations and dysfunctions, it has also inspired him: 'It is a place that makes transformation visible,' he wrote in his blog. 'Permaculture principles turn a dry wasteland into a flourishing garden. Connection to a culture and history of political movements make social change seem possible. Encountering new ideas and experiences, we begin to feel ourselves change and believe that the way things are presently need not be how they have to be.'

Despite her relative wealth, Margaret's lifestyle is simple and generous. She still lives in the Peace by Peace house in Cairns, sharing with several people who don't have anything like her 'riches': some are 'dispossessed', others are 'holding out for a better world', one she has been nursing through cancer. Sometimes all this is a stretch. In the year following the trial, she moved into an improvised living space underneath the house, cool and close to the garden. She relished having her own space down there, at least while the dry season lasted.

I watched her rehearse with the North Queensland Chamber Players. Four women, two on violin, one on cello, Margaret on viola.

They give two concerts a year, the next was in a fortnight's time. I listened, looking out from the rehearsal room to the lush green hills never far in this small city, a few tall clouds building behind them. On the rise and fall of Borodin's music, I felt the poignant sweetness yet strangeness of our European culture transplanted to this country. Margaret had struggled with this in her young adulthood, stopped playing, drifted away from her religion, but lessons she learned from her activism as well as her co-counselling practice brought her back. 'If you want to do anti-racism work, you have to root yourself in your own culture. I went back to church, it had never done me any harm, actually, it had done me a lot of good.' For one, it had instilled her love of music.

We visited St Monica's Cathedral. During the trial Margaret had mentioned her involvement with its congregation. It was where she and Bryan were married and where his funeral was held. Its bishop, James Foley, wrote to the court, vouching for her good character. Since then, though, she had stepped away from the cathedral and started going to a small neighbourhood church, St Francis Xavier. She feels a strong connection with the parish's lay leaders, 'brown women' – Aboriginal, Indian, a woman from PNG. 'If I get sick, they are the people who will come and visit me.' Meantime, what she has found there is a regular collective space for reflection. 'Where else would I get that, the time to reflect on the totality of my life?' If you don't do that, she added, 'you are likely to take the position you are given, rather than the one that you think is right'.

We drove out from Cairns to visit Paul who has joined a rural cooperative. He had a ready answer when I asked what had brought him to live on this unserviced land on the Atherton Tableland: 'The writing is on the wall, the train wreck is about to happen. Political, economic, environmental instability. It's going to be difficult for everyone to do well. That can only happen by people living together, working together.' He was brown from the sun, lean and strong-limbed from physical work. His hair, which he'd combed slick during the trial, was standing up in sun-bleached dusty tufts.

This was high, dry country, not lush like the coast. The house block was fairly bare when we visited, except for a large tin shed, which for the time being was home; alongside it was the beginning of a vegetable garden. The vision of the cooperative – then yet to join him in the day-to-day labour – is to build a large shared house there in the middle of an orchard of three thousand trees. At the time all I could see was the size of the undertaking but down by the dam – wide and full, more like a lake, water-lilies in bloom on its surface – I began to sense the allure and the potential. A solar-powered pump takes water up to the house block and will eventually irrigate their orchard and gardens. In the interim, while the co-op gets all that established, they hire out this site – with large, well-built shade structures and furnishings – and other small secluded camping areas around the property. A men's group was coming that very night. Paul talked about holding workshops there for skilling up activists. 'But you have to do that in the context of actual issues,' said Margaret. There's no shortage: Paul talked about blockading the supply runs to West Papua out of Cairns Port; about disrupting AFP training for security forces in Indonesia, involved in killing civilians in West Papua. Within the month of my visit, the Gimuy Pilgrims organised a 'pop-up concert' outside the AFP base in Cairns, drawing attention to the issue. Only a dozen or so people were involved but they had a visual impact with their Morning Star flags of the Free Papua movement and large banner reading 'West Papua AFP Shame'.

Back at the Peace by Peace house Emmaline came around. She used to live there and is a good friend but that day she was interviewing Margaret as part of a community asset mapping project for her employer, Centacare FNQ. I listened in while Emm – as Margaret calls her – plotted into her framework the many groups Margaret is involved with, the skills she has to offer, the things that are important to her. Emm often responded to Margaret's turn of phrase, like when she spoke of working for 'the right relationship between people, planet and God, the creational order of things – social justice comes out of that'. 'The creational order of things,' said Emm, 'I love that.' Margaret spoke of the importance of the 'full body experience' in actions, adapting a concept of liturgical practice from the Second Vatican Council, 'full conscious active participation'. She spoke of the 'double healing' involved in Frontier

Wars ceremonies, for Indigenous and non-Indigenous peoples – 'the only way forward, the only way to save the land'. Emm repeated the phrases reflectively.

Margaret seemed to hit a raw nerve inside herself when she talked about her ideal of 'a society that doesn't have alcohol in it'. She angrily described the devastating impact of irresponsible licensing in remote communities on the Cape – 'the sign of it is the roaming children' – and the proliferation of take-away liquor outlets in Cairns. It didn't take long to get to talk of her mother who was an alcoholic, increasingly so as she got older. Recently a woman in the neighbourhood had been stabbed to death, another one had been bashed, both from drinking households. 'I'm sick of living in a community where people kill each other because of alcohol.'

Finally Emm closed her computer. Almost three hours had passed. The two started talking as the old friends they are, about people they knew, their children, future plans. They moved into the garden, lay down side by side, Emm's long dark hair spread out on the soft cool grass, both of them gazing up into the branches of a huge paperbark tree. Margaret pointed out a pair of Torresian pigeons roosting there, then another; a willy wagtail flitted through their field of vision, then a honeyeater. Margaret showed Emm the bird app on her phone that helps her identify their calls. 'Does it have the cassowary call?' Emm wanted to know. It does, a strange vibrating sound, apparently rarely heard by humans. They listened to it several times, delighted.

Tim went back to New Zealand as soon as he'd been sentenced. He was the only Pilgrim I didn't visit at home. I suggested we talk by Skype; ever shy, he preferred email. So I sent him lists of questions and eventually got his brief but thoughtful paragraphs back.

He spent the summer after the trial at Hokianga, catching up with family, going to a wedding, enjoying a holiday. Then it was time to get serious again. He moved south to the Wellington area, living first in a Catholic Worker house in the city, called Berrigan House. 'I just want to be around these kinds of people. I can't imagine making a move anywhere that is not at least affiliated with the Catholic Worker Movement.' He enrolled at university, in religious studies, politics and history, but 'that was mostly just for fun'. The main game was getting involved in various political struggles – climate justice, blockading a weapons expo, and campaigning for the 'Hit and Run Inquiry' into alleged cover-ups by the NZ military, both at home and abroad. He and his friends would go out at night, sometimes several nights a week, doing campaign paste-ups, up to fifteen metres long, and stencilled graffiti. They also did many early morning stints leafletting outside Defence Forces HQ in the middle of Wellington, calling for whistleblowers to come forward.

The following year he moved back to the country, to the Otaki Catholic Worker farm, about an hour's drive north of Wellington. Sarah, whom he'd met at Berrigan House, had begun featuring in his emails, first as a girlfriend then as a fiancée. As the community guideline is that one or both people in a couple who become romantically involved should move out, Tim went to Otaki while Sarah stayed on to finish her studies. After that they planned to marry and build a mobile 'Tiny Home' to have a place of their own, based in Otaki initially. 'She's relatively new to activism,' Tim wrote, 'but yes, very interested in it all. She's about to graduate a BA with a double major in anthropology and religious studies and a minor in Maori studies. Just to give you an idea of the way she is approaching the subject of activism. So she understands me!' He was referring to the way he sees his activism as an extension of his Christian commitment to loving his neighbour: 'In a time when everything is so globalised our "neighbour" has come to also encompass the rest of the world basically. I believe that part of loving those neighbours takes the form of standing up to their oppressors, such as those in the Pine Gap Joint Defence Facility.'

~

I was drawn to write about the Peace Pilgrims because of the large view they take of their social responsibilities. They may join campaigns, but really, their field of action is the whole of life, as far as their capacities and nonviolence can take them. They are connected to movements, and even specific groups within them, but within the bounds of their strong spiritual and moral frameworks they seem remarkably free – unconstrained by waiting for consensus, or theoretical coherence, or numeric strength, or likely success. In this way what they do is very open, an offering, for others to make of what they will.

The Commonwealth's oppressive prosecution tried to make an example of them. At the time of writing this approach to dissent was being ramped up with the prosecution of whistleblowers and attacks on press freedom, particularly in relation to reporting on 'national security' matters. It remains to be seen how this will play out in Australia's political culture in the longer term. Many prominent Australians rallied behind the Pilgrims – the pro bono legal support, the signatories to the open letter to Brandis, scores of letters asking for clemency that were sent to Justice Reeves. A significant part of that support was around civil liberties issues; that was the emphasis in the open letter to Brandis, and in the appeals for lenient sentencing by prominent lawyers. Excessive penalties would violate the rights to freedom of expression and assembly, it was argued, rights guaranteed by Australia's international commitments and the implied right to political communication under our Constitution. Reeves avoided being excessively punitive, but there remains the fact of the Crown resorting to prosecution under the *DSU Act* in the first place: that required a political decision, and what resulted was a political trial. The Pilgrims, to the extent they were able, fought fire with fire. McHugh had argued that it came 'close to an abuse of the court's processes', but the Commonwealth went there first and kept it up, every step of the way. These things – questions of rights, of the way power is used – matter to the Pilgrims but they are not the core of what motivates them. They know their desire for peace, their rejection of war and militarism ultimately need the public square more than the court, filled by people in a mass movement. And they need the floor of parliament, filled by political representatives who hold themselves accountable to the people they serve.

EPILOGUE

The Pilgrims' trials, the sentencing, were two summers past when I took a break from the writing, spending the last days of January in 2019 camping with Erwin in sand dunes near Perlubie on the Eyre Peninsula. Away from farming land, the wild country there felt as remote as the deserts of central Australia. The dunes were hot and implacable at that time of year; for all their beauty it was arduous walking across them to the sea. The dazzling beach, covered with white pipi shells, made our eyes ache. We'd throw down our packs and run to the shallow bay waters, their clear cold relief. Occasionally a fishing boat would pass; once or twice, a cargo ship on the horizon. Sometimes, on the way back through the dunes, we'd lose our way. I had a sense of hanging off the edge of the world. In the town of Streaky Bay, where we went for supplies, we saw just one sign of outside worries intruding: a mural protesting proposed deep sea oil drilling in the Bight.

With no after-dinner television, the star-filled nights were our spectacle. As we watched, more and more stars became visible, none bigger and brighter in the January skies than Orion, the hunter of Greek mythology, with sword and shield, and club raised, in relentless pursuit of the Pleiades, the Seven Sisters. I complained to Erwin about this story of predation – why do we have to see it this way? I tried to shake it from my vision, but just as with the Emu, once seen, it is hard to unsee. Unless something comes in its place.

I was staying in touch with my book indirectly, lying in the camp in the afternoon shade, listening to a podcast series about war, 'The Mark of Cain', by the Canadian historian Margaret MacMillan. While broadly arguing for a much greater awareness of how easy it is to get into wars and how hard to get out of them, especially in ways that do not provide the seed bed for future conflict, she also proposed that war is part of human nature. Hence her title. The lectures were given to live audiences in different venues. In one, she was asked by a woman if she thought we, humankind, would be as warlike if women were in charge. The audience laughed and clapped – what an entertaining idea! MacMillan looked to history for her answer: to the Spartan mothers of the Peloponnesian Wars, whose message to their sons was 'Come back bearing your shield or on it', in other words, victorious or dead; to women of the nineteenth century such as Bismarck's wife, who expressed herself in ferocious terms about what should be done to the enemy; to the women leaders of their nations in the twentieth century, Indira Gandhi, Sirimavo Bandaranaike, Golda Meir, Margaret Thatcher. They did not seem to be any more peaceable than male leaders, she observed.

I felt deeply dissatisfied with this answer. Not just because of my own anti-war sentiment or wishful thinking, but because of how reductive MacMillan's argument is. Her examples, obviously, are all of women formed in patriarchal societies – who among us are not? These women expressed themselves and acted within the power structures and social norms that had shaped them and contained them still. In the case of the national leaders, rare exceptions in gender terms, they had responded to the existing leadership mould and competed successfully for the position, rather than challenging in any fundamental way the rules of the game.

A very differently conceived work, *A Terrible Love of War* by the Jungian psychologist James Hillman, also holds that war is fundamental to humans, 'an indelible condition of the soul'. He too dismisses any intrinsic difference between men and women in this regard. To counter what he calls 'the testosterone hypothesis' of war, he looks to the legendary Amazons, as well as, like MacMillan, to modern women in power who have been war leaders, and others who have clamoured for entry into military life and have excelled as soldiers. Patriarchy, he contends, rather than being the originator of war, is its necessary result, 'preventing Ares [Greek god of War] from blowing up the world'. Elsewhere in the book, Hillman embraces approaching 'the unknown by way of unknowing' and he is wide-ranging, erudite and imaginative in doing that, but here, it seems to me, he is merely falling back on conventional wisdom.

I am frustrated too by his hollow idea of peace, that it is, at best, an absence of war, a breathing space between wars, at worst, a cover-up, bloated with idealisations. He would never march for peace, nor pray for it, which brings his fighting words right into the territory of this book. Humanity is mortality, he writes in his pages 'thick with death' – we all die, always have, always will. He skips over the corollary, indeed the precondition – we are all born, and therein peace finds its substance. Peace is 'the nurture of human life', as Jane Addams said. She began her peacework during the First World War and was awarded the Nobel Peace Prize in 1932. Nothing idealised in her formula. Nurture is work that can fill all our days at every level, from hearth to planet, and never more urgently than now.

In our camp in the dunes, I was also proofreading my daughter Jacqueline's doctoral thesis (I have to laugh – a real busman's holiday). She was looking at ways the maternal is represented in art and poetry, drawing on new understandings from feminist psychoanalysis about subjectivity – what is signified when we say 'I'. These ideas re-conceive the subject as forming not with severance from her mother – as psychoanalysis in the Freudian tradition would have it – but in relation with her from the beginning, in the uterus, through birth, and on into the world, always relational, continuous and embodied. This is such a different notion to the violence of 'the cut' as the starting point for subjectivity, and its supposed necessary evolution through repression of the relation with the mother and entry into the hierarchical law of the father. What could happen if we understood and valued a structure and process for subjectivity that was inclusive and generative? Women can intuit something of this surviving within and between themselves, but historically and culturally it has been subordinated when not utterly erased.

I'm collapsing a whole complex field of inquiry into a few sentences here, but in my summer reading Jacquie's thesis helped me talk back to MacMillan, to Hillman. Can we know, I wanted to ask them, what kind of course humanity, differently formed, would take? Would subjects formed with reciprocity at their valued core not generate a very different kind of social order? Creating the conditions of peace, ending war, would seem to require nothing less. For who can feel confident right now in patriarchy's ability to prevent Ares from 'blowing up the world', whether that's by nuclear catastrophe or in the failure to deal with climate change?

Erwin and I headed back to Alice Springs in early February, taking the back road from the Eyre Highway through the Gawler Ranges. It was very hot, forty plus. Willy-willies whipped across the ground, the flies were intense, all we could do was drive. The country in the broad valley we were passing through looked greener than I'd expected and I was surprised to see, among other trees, tall native pines, the kind that grow at Kweyernpe and along the ranges in the Centre. Suddenly, a commotion: an emu was under our wheels. It had burst from thick scrub right at the edge of the road. Erwin got out: the emu was dead. I stayed in the car, not wanting to see. We were quiet for a while after that; we hadn't been going particularly fast, eighty, maybe ninety, but we felt a kind of shame.

We made our camp that night in a clearing off the Stuart Highway. At first light, before the flies woke, I went for a walk along the fence-line. An emu had been there before me. I followed the tracks in the red sand for a kilometre or so, wondering where and if this bird had made it through to the bush on the other side.

For the rest of the morning we were travelling through the Woomera Prohibited Area (WPA), which stretches from Woomera in the south to just shy of Coober Pedy on the eastern side of the highway, as far north as Cadney Park on the western side, more than 100,000 square kilometres of land. We'd driven this route for years, noticed the warning signs by the side of the road and thought of them as a chilling relic of the British nuclear tests at Maralinga. Not at all. After a decline in the 1980s and '90s, the WPA has resumed its role as an active weapons testing range, the largest in the world, ideal for its remoteness and 'quiet electromagnetic environment', open for business.

Again it was Kristian Laemmle-Ruff who brought this to my attention. He had joined a tour of Australia's nuclear landscapes, run each year by Friends of the Earth; one of their stops was Woomera. He later went back with a couple of friends and camped on the shores of Lake Hart at the southern end of the prohibited area. Erwin and I have often stopped there, at least to stretch our legs, walking down from the road to where the desert sand meets the glittering salt crust of the lake, staining it rust and ochre. Kristian and his friends went a bit further, penetrating into the prohibited area's red zone, exclusively for Defence use year round. When he wanted to go even deeper into the zone, onto a part of the area known as Evetts Field, his friends left him to it. The whole area is governed by the *DSU Act* – the same legislation used to prosecute the Pilgrims – so once again just by being there he was exposing himself to repercussions.

I can remember the naive excitement with which I ventured up this highway for the first time more than thirty years ago, into the big landscape, the tiny rough-edged communities, the sense I had of openness and possibilities. Now I feel the weight of the creeping expansion of a militarised swathe running south to north right through the centre of Australia. RAAF Base Edinburgh, Technology Park in the north of Adelaide – both subject to massive new investment by the Commonwealth, Lockheed Martin, Raytheon, BAE Systems, SAAB – up through the WPA, parts of it in near constant Defence use, to the Pine Gap military base, going through a 'midlife upgrade' for an unquantified spend, and Jindalee Operational Radar Network in the Centre, part of a $1.2 billion upgrade, to RAAF bases in Tindal and Darwin, barracks in Palmerston and Darwin, expansion of the United States Force Posture Initiative training area in Darwin, the Australian Signals Directorate facility at Shoal Bay nearby, and a possible new deep-water port development a little further out, offering an improved base for the regular 'rotations' of US Marines and their deployment into the Indo-Pacific. The language used to talk to the public about all this is expressed in terms of industry, jobs, investment, billions of dollars. Nothing about its fundamental premise of violence and threat as the way to conduct human affairs.

There is no transparency around the kind of weapons tested in the WPA, but one example the ABC uncovered is the British-built Taranis Unmanned Combat Air Vehicle – the drone nicknamed 'Raptor' – designed to carry a payload of guided missiles and bombs. At the time of the test, 2013, the British military claimed it would be able to operate the Taranis via satellite link from any location in the world, and, like the US B-2 stealth bombers, it would be able to fly undetected by radar.

The images Kristian made of the WPA show no weapons, no explosions, no craters. The idea of them is evoked instead, with elegant simplicity, by a laser beam. Each image was made in one photographic exposure, taken at night, with the shutter wide open and Kristian running in a circle on the salt lake, using a flash to light its surface and switching on the laser from different points around the periphery. All this in a military zone where he had no permission to be.

At each intersection of the laser beams is the idea of target, and in the imagined aerial space overhead, the idea of all the links – human and machine – in the kill chain. What is targeted, in the images, is the earth, closely intimate, standing in for all of us – for without it we would soon be nothing. The desert is often thought of as empty, even dead and deadly, but Kristian's imagery here shows it deeply connected with life – seen in the welling-up of moisture making circular plaques across the skin of the lake, and in the imprint of animal life, the tracks of the emu.

Franz Dowling, Margaret Pestorius, Jim Dowling, Andy Paine and Tim Webb, Alice Springs, 16 November 2017. (Photo by Cate Adams)

Paul Christie and Graeme Dunstan, Alice Springs, 13 November 2017. (Photo by Kieran Finnane)

NOTES

PROLOGUE

p. vii 'The Queen's crown': Quoted in Kieran Finnane, 'Their share of the struggle: innovation & experimentation in desert arts', *Art Monthly* Australia, issue 260, June 2013. For his essay expanded from the keynote, see Wanta Steve Patrick Jampijinpa, 'Pulya-ranyi: Winds of Change', *Cultural Studies Review*, vol. 21, no. 1, March 2015, pp. 121–31, accessed online 2 May 2018.

p. viii Collated sources about this drone strike in Jessica Purkiss, 'Afghanistan: Reported US covert actions 2017', The Bureau of Investigative Journalism, entry AFG269, accessed online 26 April 2018.

Authorised by Obama: Josh Smith, 'US drone strike kills 15 civilians in Afghanistan, United Nations says', Reuters, 30 September 2016, accessed online 26 April 2018.

Raghon Shinwari quoted by Rafiq Sherzad, Ahmad Sultan, Josh Smith, 'US strike on Islamic State in Afghanistan kills 21, maybe some civilians', Reuters, 28 September 2016, accessed online 26 April 2018.

Village named as Shadal Bazar; quotes from Obaidullah, Mohabad Khan: Sune Engel Rasmussen, 'Deadly US drone strike in Afghanistan not disputed but victims' identities are', *Guardian*, 1 October 2016, accessed online 23 April 2018.

United Nations Assistance Mission in Afghanistan, 'UNAMA condemns killing of at least 15 civilians in airstrike', press release, 29 September 2016, accessed online 26 April 2018.

US rationale for attack: USFOR-A updated statement, 1 October 2016, accessed online 23 April 2018.

International law protecting civilians: Timothy LH McCormack and Helen Durham, 'Aerial bombardment of civilians: The current international framework', in Yuki Tanaka and Marilyn B Young (eds), *Bombing Civilians*, The New Press, New York, 2009, pp. 220–5.

International legal obligation on states to report in detail on civilian injury and/or death: Ben Emmerson, Special Rapporteur, 'Promotion and protection of human rights and fundamental freedoms while countering terrorism', report to United Nations General Assembly, A/68/389, 18 September 2013, paragraph 78.
'Not a civilian casualty incident': email, Press Desk, Resolute Support Public Affairs, Kabul, Afghanistan, 8 June 2018, replying to my emailed enquiry, 24 May 2018.
Haji Rais list: Purkiss, op. cit.

SEEING, NOT SEEING

p. 1 Kristian Laemmle-Ruff, *Pine Gap (a photograph of the Centre of Australia)*, 2015. The work was shown in the Alice Prize at the Araluen Arts Centre in April 2016 and in a solo show, *Mind the Gap*, at Watch This Space, Alice Springs, in September 2016. The account of his experience in making it: personal communication, 19 May 2018.

p. 2 ICAN: The name and acronym are owed to Ron McCoy, Malaysian obstetrician and former co-president of International Physicians for the Prevention of Nuclear War (IPPNW): see Tilman Ruff, 'How Melbourne activists launched a campaign for nuclear disarmament and won a Nobel prize', *Conversation*, 9 October 2017, accessed online 19 May 2018.
Australia and the ban treaty: AAP, 'Nuclear ban treaty agreed despite boycott by US, UK and other major powers', *Sydney Morning Herald*, 8 July 2017, accessed online 11 June 2018; Michael Slezak, 'Australia attempts to derail UN plan to ban nuclear weapons', *Guardian*, 21 August 2016, accessed online 11 June 2018; Ben Doherty, 'UN votes to start negotiating treaty to ban nuclear weapons', *Guardian*, 28 October 2016, accessed online 11 June 2018.
ICAN awarded Nobel Peace Prize: 'The Nobel Peace Prize 2017', Nobel Media, accessed online 11 June 2018.
Richard Tanter is Honorary Professor in the School of Political and Social Sciences at the University of Melbourne and a Senior Research Associate of the Nautilus Institute for Security and Sustainability.

p. 3 'An increasingly comprehensive Five Eyes surveillance system': Desmond Ball, Duncan Campbell, Bill Robinson and Richard Tanter, 'Expanded communications satellite surveillance and intelligence activities utilising multi-beam antenna systems,' NAPSNet Special Report, 28 May 2015, accessed online 17 August 2018. Tanter's photo of the Torus is published on p 32. The report also discusses Torus antennas operating in Russia and Ukraine, and others owned by the US.
Comments on their work: Richard Tanter, personal communication, 18 December 2018.

p. 4 Pine Gap's technology and its functions: Desmond Ball, Bill Robinson and Richard Tanter, 'The antennas of Pine Gap', NAPSNet Special Report, 22 February 2016, accessed online 22 May 2018; Richard Tanter, 'Our poisoned heart: The transformation of Pine Gap', *Arena Magazine*, no. 144, Oct/Nov 2016, pp. 12–14.

p. 6 Targeted killing outside zones of armed conflict: Philip Alston, Special Rapporteur, 'Study on targeted killings', submission to United Nations

General Assembly, Human Rights Council, A/HRC/14/24/Add.6, 28 May 2010, paragraph 33. Signature targeting: Grégoire Chamayou, *Drone Theory*, translated from the French by Janet Lloyd, Penguin Books UK, London, 2015, ch. 5, 'Pattern-of-life analysis'. Criteria for legitimate targets: Christ of Heyns, Special Rapporteur, 'Extrajudicial, summary or arbitrary executions', report to United Nations General Assembly, A/68/382, 13 September 2013, paragraph 72; see also Emmerson, op. cit., paragraph 74. 'Near certainty': 'US policy standards and procedures for the use of force in counterterrorism operations outside the United States and areas of active hostilities', 23 May 2013, Obama White House archives, accessed online 27 June 2018. Extrajudicial assassinations: Alston, op. cit., paragraphs 33, 70, 85. Relative to the standard of 'beyond reasonable doubt': Nick McDonell, 'The targeting and killing of a Helmandi combatant', *Longreads*, 1 October 2018, excerpt from *The Bodies in Person: An account of civilian casualties in American wars*, Blue Rider Press, New York, September 2018, accessed online 5 October 2018.

Likelihood of Pine Gap involvement in Shadal Bazar strike: Richard Tanter, personal communication, 18 December 2018.

p 8 Town Council policy: Erwin Chlanda, 'Council's one sentence Pine Gap policy is 30 years old', *Alice Springs News*, 7 October 2016, accessed online 13 June 2018; see also council's website, policy 111.

'Blank spots on the map': Trevor Paglen's photograph, *An English Landscape (American Surveillance Base near Harrowgate, Yorkshire)*, was installed at the Gloucester Road tube station for the Art on the Underground programme, 15 June 2014 – 13 July 2016, archive accessed online 17 August 2018. 'Blank spots' refers to the title of Paglen's book *Blank Spots on the Map: The Dark Geography of the Pentagon's Secret World*, Dutton Adult, New York, 2009. See also Diane Smyth, 'An English landscape', *British Journal of Photography*, 19 June 2014, accessed online 17 August 2018; and Maev Kennedy, 'The intelligence of art: Yorkshire surveillance centre put on display', *Guardian*, 16 June 2014, accessed online 15 August 2018.

Pine Gap needs images: Tanter, personal communication, 15 August 2018. See also his essay, 'Landscapes of secret power: Pine Gap and Menwith Hill', *Arena Magazine*, no. 147, April 2017, pp. 26–30, with photography by Kristian Laemmle-Ruff, Trevor Paglen, Felicity Ruby and Desmond Ball, accessed online 18 August 2018.

p. 9 Kristian's photographs published in Ball, Robinson and Tanter, 'The antennas of Pine Gap', op. cit.

'Arresting' images: John Berger, 'Photographs of agony', in *About Looking*, Pantheon Books, New York, 1980, pp. 37–40.

'A radial system': John Berger, 'Uses of photography', in *About Looking*, op. cit, pp. 48–63.

p. 10 'A terror facility': Kristian Laemmle-Ruff, 'Mind the Gap', exhibition catalogue, 2016, p. 6.

'How I began my speech': opening remarks, slightly edited, Kieran Finnane, 'Pine Gap: wake-up call for Alice Springs', *Alice Springs News*, 17 September 2016, accessed online 11 June 2018.

p. 11 'Teaching viewers how to see or not see': Sarah Sentilles, *Draw Your Weapons*, Text, Melbourne, 2017, p. 271.

'The wars my country fights': Sentilles, ibid., p. 161.

p. 11 'The world is made': Sentilles, ibid., p. 269.

p. 12 'Opposed broadly to the presence of US forces': See 'The IPAN Statement' on their website, accessed 11 June 2018.
'Cycles of terrorism in response': Tanter in IPAN, 'Pine Gap a threat to Australian security', media release, 15 September 2016.
'Most serious security breach': David Rosenberg, *Inside Pine Gap: The spy who came in from the desert*, Hardie Grant Books, Melbourne, 2011, pp. 145–6.
Details of trespass: Erwin Chlanda, 'The picture they didn't want you to see', *Alice Springs News*, 21 December 2005.

p. 13 'A liveable life and a grievable death': Judith Butler, *Precarious Life: The Powers of Mourning and Violence*, Verso, London and New York, 2004, p. xv.

p. 14 'Violence … against those who are unreal': Butler, ibid., p. 33.

IN THE DARK

p. 15 Surveillance at Kweyernpe: Erwin Chlanda, 'Spying on park visitors: nothing to see here?', *Alice Springs News*, 26 July 2019. Stone tools, grass seed etc. at Kweyernpe: Dr Fiona Walsh, personal communication, 13 June 2019.

p. 17 Fiftieth anniversary: Erwin Chlanda, 'Stealth party to mark 50 years of Pine Gap', *Alice Springs News*, 24 July 2017.
Convergence protests: Erwin Chlanda, 'Pine Gap demonstrators provide real time images', *Alice Springs News*, 29 September 2016.

p. 18 'Killed us with kindness': Felicity Ruby, 'Minding the Gap: Peace Pilgrims' face seven years' jail for protesting Pine Gap's global surveillance capabilities', *Arena Magazine*, No. 149, Aug/Sep 2017, pp. 8–10, accessed online 1 March 2019.
'Consistent life ethic': Joseph Bernardin, 'A consistent ethic of life: An American-Catholic dialogue', Gannon Lecture, Fordham University, 6 December 1983, accessed online 6 July 2018.

p. 19 Catholic Worker Movement: James Allaire and Rosemary Broughton, 'An introduction to the life and spirituality of Dorothy Day', The Catholic Worker Movement website, accessed 3 March 2018. 'The aims and means of the Catholic Worker': reprinted from *The Catholic Worker* newspaper, May 2018, 85th Anniversary Issue, on The Catholic Worker Movement website, accessed 3 March 2018. Dorothy Day and abortion: Jim Dowling, personal communication, 4 December 2019. Opposition 'quiet and personal': Brian Terrell, 'Dorothy Day: "We are not going into the subject of birth control at all as a matter of fact"', *National Catholic Reporter*, 30 September 2015, accessed online 20 November 2019. Nurturing life requires sacrifice: Alice Lange, 'Dorothy Day on women's right to choose', *Houston Catholic Worker*, 1 June 2008, accessed online 31 December 2019.

p. 20 Franz Jägerstätter: Canon Paul Oestreicher, 'Face to faith,' *Guardian*, 21 October 2007, accessed online 26 July 2018; Andrew Hamilton, 'Jägerstätter, a man who acted on conscience', *Eureka Street*, 6 June 2007, accessed online 26 July 2018. Jägerstätter's life is also told in a beautiful film, *Der Fall Jägerstätter* (released as *The Refusal* in English), directed by Axel Corti, 1972.

p. 24 ANZAC Waihopai Ploughshares, on Lest We Forget: Remembering Peacemakers on Anzac Day website, accessed 25 July 2018.

p. 27 Joint Chiefs of Staff, United States Department of Defense, 'Joint Concept for

Integrated Campaigning', 16 March 2018, accessed online 3 October 2018.

p. 28 Nano explosives, 3D printing of drones, swarms, AI improvements, role for older tech missiles: Dr Thomas X Hammes, Institute for National Strategic Studies, National Defense University, 'Uninhabited Aerial Systems: Disruption or Prescription?', paper presented to 2018 Air Power Conference, *Air Power in a Disruptive World*, 20–21 March 2018, National Convention Centre Canberra, proceedings archived online, accessed 8 October 2018. In late 2018 the Australian Defence Force acquired its own armed drones, General Atomics MQ-9 Reapers: relatively large and expensive (at an estimated $23 million each) and carrying much heavier weapons payloads. See 'General Atomics Reaper selected for Australia's first armed remotely piloted aircraft system', joint media release, Minister for Defence, Christopher Pyne and Minister for Defence Industry, Steven Ciobo, 16 November 2018, accessed online 1 January 2019. On the number and expense: David Wroe, 'Australia to buy armed Reaper drones in shift towards pilotless future', *Sydney Morning Herald*, 15 November 2018, accessed online 1 January 2019. 'Top 10 weapons export league': Gareth Hutchens, 'Australia unveils plan to become one of world's top 10 arms exporters', *Guardian*, 29 January 2018, accessed online 13 July 2018. Government's 'heavy' investment: Marise Payne, Minister's Address, 2018 Air Power Conference, op. cit. '$195 billion': Department of Defence, '2016 Defence White Paper', 25 February 2016, accessed online 15 June 2018. AI as our 'intellectual partner'; our ethics as 'handcuffs': JD McCreary, Chief Disruptive Technology Programs, Georgia Tech Research Institute, 'The human/machine interface: Operational decision-making', 2018 Air Power Conference, op. cit. Hybrid warfare used for good: Peter Jennings, Australian Strategic Policy Institute, 'Thriving or just surviving? Australia's tough choices in a risky strategic age', 2018 Air Power Conference, op. cit.

p. 29 Obama on nuclear disarmament: Dan Zak, *Almighty: Courage, resistance, and existential peril in the nuclear age*, Blue Rider Press, New York, 2016, pp. 154–7. His Nobel: 'The Nobel Peace Prize for 2009 to President Barack Obama – press release', Nobel Media, 9 October 2009, accessed online 7 June 2018. He entrenched the US nuclear arsenal for decades: Zak, op. cit. pp. 292–3. His record for secret killing of individuals: see, for example, Greg Miller, 'Under Obama, an emerging global apparatus for drone killing', *Washington Post*, 27 December 2011, accessed online 11 June 2018.

p. 30 *Authorisation for the Use of Military Force*: S. J. Res. 23, 107th Congress of the United States of America, accessed online 27 June 2018. It was followed in 2002 by a specific AUMF against Iraq: Public Law 107–243, joint resolution, 107th Congress, 16 October 2002, accessed online 3 January 2019. Lethal force permissible 'strictly to prevent the imminent loss of life': Alston, op. cit., paragraph 93. Anticipatory self-defence, legal test: Alston, op. cit., paragraph 86. Heyns, op. cit., paragraph 87. Battlefield, the world – assault on territorial integrity. Chamayou, op. cit., p. 53.

Unlimited presidential power: Malcolm Fraser with Cain Roberts, *Dangerous Allies*, Melbourne University Press, 2014, p. 205. Fraser's views had cut-through: see, for example, Graeme Dobell, Peter Jennings and Peter Edwards, 'Reassessing Malcolm Fraser', Australian Strategic Policy Institute, accessed online 29 May 2019. 'We need the United States for defence': Fraser with Roberts, op. cit., Introduction.

p. 31 Obama's use of AUMF: 'Legality of drone warfare', Explainer, The Bureau of Investigative Journalism, accessed online 23 April 2018. See also Rachel Stohl, 'An Action Plan on US Drone Policy: Recommendations for the Trump Administration', The Stimson Center, 7 June 2018, full report, accessed online 11 June 2018. Strikes during Obama's terms: Jessica Purkiss, 'Obama's covert drone war in numbers: ten times more strikes than Bush', The Bureau of Investigative Journalism, January 17, 2017, accessed online 21 May 2018. Civilian casualties estimates: The Bureau of Investigative Journalism logs put the figure between 384 and 807, op.cit.

Drone strikes in Trump's first year: Jessica Purkiss, 'Trump's first year in numbers: strikes triple in Yemen and Somalia', The Bureau of Investigative Journalism, 19 January 2018, accessed online 21 May 2018. This article also provides estimates for Afghanistan. Reported strikes in 2019, timelines for Afghanistan, Pakistan (no reported strikes), Somalia and Yemen: ibid., accessed online 1 February 2020. Trump handing off oversight to the Pentagon: Julian Borger, 'US air wars under Trump: increasingly indiscriminate, increasingly opaque', *Guardian*, 23 January 2018, accessed online 11 June 2018; Zachary Cohen and Dan Merica, 'Trump takes credit for ISIS "giving up"', CNN, 17 October 2017, accessed online 11 June 2018; Jessica Purkiss and Abigail Fielding-Smith, 'US air strike data from Afghanistan takes step back in transparency', The Bureau of Investigative Journalism, 20 December 2018, accessed online 3 January 2019.

History of bombing and civilian casualties: Mark Selden, 'A forgotten Holocaust: US bombing strategy, the destruction of Japanese cities, and the American way of war, from the Pacific War to Iraq', in Tanaka and Young (eds), op. cit., pp. 77–96. 'Customary law' of airwars: Sven Lindqvist, *A History of Bombing*, translated by Linda Haverty Rugg, The New Press, New York, 2001, p. 113, section 239. Legal tests for use of weaponised drones, same as for other weapons: Alston, op. cit., paragraph 79. War crimes: McCormack and Durham in Tanaka and Young (eds), op. cit., pp. 220–5. 'Diffusion of responsibility': ibid., p. 217.

Tagore on the aeroplane and early aerial bombardment: Barry Hill, *Peacemongers*, UQP, St Lucia, Queensland, 2014, pp. 70–2.

p. 32 Bombing 'barbarians', utility in the colonies, international law: Lindqvist, op. cit., p. 2, section 5, pp. 42–3, section 102, pp. 49–50, sections 115–17. Europeans bombed themselves: Selden, op. cit., pp. 80–2; Lindqvist, op. cit., pp. 82–3, section 179. Aerial bombing excluded from Nuremberg and Tokyo trials: Selden, op. cit., p. 92. Allies worked to limit Geneva Conventions: Lindqvist, op. cit., pp. 121–2, section 256. The American 'way of war' and elusive victory: Selden, op. cit., pp. 92, 96.

Drones lower threshold for resort to force; outside armed conflict, almost 'never likely to be legal': Alston, op. cit., paragraph 80. Obligations of transparency and accountability: Alston, op. cit., paragraphs 87–92; Heyns, op. cit., paragraphs 108–10; Emmerson, op. cit., paragraphs 41–5. In his submission to the 2015 Senate Committee, 'The potential use by the Australian Defence Force of unmanned air, maritime and land platforms' (accessed online 20 February 2019), Ben Saul, Professor of International Law, Sydney Centre for International Law, quotes from UN Human Rights Council resolution 25/22 (adopted 28 March 2014), which calls upon states 'to ensure transparency in their records on the use of remotely piloted aircraft or armed drones and to conduct prompt, independent and impartial investigations whenever there are indications of a violation to international law caused by their use'. Reparations or remedies must be provided in cases where the right to life has been violated.

p. 33 This 'different kind of war': 'President Obama Speaks on the US Counterterrorism Strategy' at the National Defense University', transcript, Obama White House Archives, 23 May 2013, accessed online 12 June 2018. Man-hunting, 'depriving the enemy of an enemy', the warrior ethos troubled, soldier a 'mere executioner', psychic costs: Chamayou, op. cit., ch. 3, 'The theoretical principles of man-hunting', ch. 7, 'Counterinsurgency from the air', ch. 11, 'A crisis in military ethos', ch. 12, 'Psychopathologies of the drone'. Policing operation without other tools of policing, Heyns, op. cit., paragraph 103. Asymmetry: McCormack and Durham in Tanaka and Young (eds), op. cit., pp. 236–7.

Claims of surgical precision: see, for example, Miller, op. cit.; for an Australian example, see John Blaxland, 'Pine Gap at 50: why controversy lingers and why its utility is enduring', *The Strategist*, Australian Strategic Policy Institute, 21 August 2017, accessed online 31 May 2018. For a challenge to the claim of precision, see Larry Lewis, a principal research scientist at the Center for Naval Analyses, quoted in Spencer Ackerman, 'US drone strikes more deadly to Afghan civilians than manned aircraft – adviser', *Guardian*, 2 July 2013, accessed online 15 June 2018.

'New desire for revenge': David Kilcullen and Andrew McDonald Exum, 'Death From Above, Outrage Down Below', *New York Times*, 16 May 2009, accessed online 2 July 2018. British officer in 1923 on aerial bombardment in Iraq: Chamayou, op. cit., p. 63.

p. 34 Pine Gap's SIGINT contribution to war: Tanter, personal communication, 24 June 2019. Profiles of Pine Gap personnel: Desmond Ball, Bill Robinson and Richard Tanter, 'The militarisation of Pine Gap: Organisations and personnel', Nautilus Institute for Security and Sustainability Special Report, 14 August 2015, p. 15 & fn 21, p. 21 & fn 34 , pp. 24–5 & fn 48, p. 28 & fn 76, accessed online 21 May 2018.

p. 35 Military context of Rosenberg's work: Rosenberg, op. cit., pp. 85, 87, 103, 110, 111.

'Whatever assets it has in place': Rosenberg in Dylan Welch, 'Top intelligence analyst slams Pine Gap's role in American drone strikes', transcript, ABC, *7.30 Report*, 13 August 2014, accessed online 13 June 2018. The intelligence analyst referred to in the headline is Desmond Ball, not Rosenberg. See also Philip Dorling, 'Pine Gap drives US drone kills', *Sydney Morning Herald*, 21 July 2013, accessed online 23 April 2018.

Cian Westmoreland: in Peter Cronau, 'The Base: Pine Gap's role in US warfighting', *Background Briefing*, ABC Radio National, 20 August 2017, accessed online 24 May 2018.

Des Ball on attack in Yemen and 'hunting down terrorists': Paul Maley, 'Pine

Gap "supports US drone hits"', *Australian*, 20 May 2014, accessed online 5 June 2018. See also 'Yemen: Reported US Covert Actions 2001–2011', entry YEM001, The Bureau of Investigative Journalism, accessed online 5 June 2018.

p 36 Ball's earlier support for Pine Gap: Transcript of his testimony to the Joint Standing Committee on Treaties, Canberra, 9 August 1999, uncorrected proof, TR 3, TR 8, accessed online 24 June 2019. See also Richard Tanter, 'The "Joint Facilities" revisited – Desmond Ball, democratic debate on security, and the human interest', Nautilus Institute for Security and Sustainability Special Report, 12 December 2012, pp. 3, 6, 43–5, accessed online 31 May 2018.

'Pine Gap doing things very, very difficult to justify': Ball in Welch, op. cit. Deaths of two Australian citizens in Yemen: 'Yemen: Reported US Covert Actions 2013', entry YEM141, The Bureau of Investigative Journalism, accessed online 31 May 2018. 'These people were with that target': Foreign Minister Julie Bishop in Paul Maley, 'ASIO saw Yemen drone pair as a risk', *Australian*, 16 May 2014, accessed online 5 June 2018. Parke on Yemen attack: Melissa Parke, Parliament of Australia Hansard, Adjournments, Thursday, 5 June 2014, accessed online 22 May 2018. Parke's CV: Biography, Hon. Melissa Parke MP, Parliament of Australia, accessed online 12 June 2018. After leaving parliament, Parke became one of three members of a Group of Eminent Experts on Yemen established by the UN Human Rights Commission in December 2017: Office of the UN High Commissioner for Human Rights, 'Yemen: Zeid appoints group of eminent international and regional experts', accessed online 17 August 2018.

p. 37 Ludlam–Brandis exchange on legality of drone strikes: Parliament of Australia, Hansard, Legal and Constitutional Affairs Legislation Committee, Estimates, Attorney-General's Portfolio, 28 May 2014, accessed online 12 June 2018. Ludlam–Brandis exchange on role of Pine Gap in strikes: Parliament of Australia, Hansard, Legal and Constitutional Affairs Legislation Committee, Estimates, Attorney-General's Portfolio, Australian Security Intelligence Organisation, 29 May 2014, accessed online 12 June 2018. Leaked Snowden documents published in Cronau, op. cit.

p. 38 Smith on 'full knowledge and concurrence': Stephen Smith, 'Ministerial statement, full knowledge and concurrence', Parliament of Australia, House Hansard, 26 June 2013, accessed online 22 May 2018.

p. 39 Fraser on 'full knowledge and concurrence': Fraser with Roberts, op. cit., p. 253. AUMF 'tenuous', Australians at Pine Gap 'equally liable': Fraser with Roberts, op. cit., p. 269. US attitudes towards the ICC: Ishaan Tharoor, 'The White House's new attack on the international system', *Washington Post*, Today's WorldView, by email, 11 September 2018. See also 'US opposition to the International Criminal Court', American Bar Association, 23 October 2012, accessed online 26 December 2018. Australia a state party to the Rome Statute: 'Support for the International Criminal Court', Attorney-General's Department, International Relations, accessed online 26 December 2018.

p. 40 Human rights advocacy groups, international law experts: see submissions to 2015 Senate Committee: 'The potential use by the Australian Defence Force of unmanned air, maritime and land platforms,', accessed online 20 February

2019.

Ministerial Statement, Australia–United States Joint Facilities: House Hansard, 20 February 2019, accessed online 20 February 2019, p. 14,049; Shadow Minister in reply: p. 14,054. 'Rather coy': Peter Jennings, 'The joint facilities: still the jewel in the crown', *The Strategist*, Australian Strategic Policy Institute, 21 Feb 2019, accessed online 29 May 2019.

p. 41 Questions to the Minister: Kieran Finnane, 'Liberals and Labor: Tweedledum and Tweedledee on Pine Gap', *Alice Springs News*, 27 February 2019, accessed online 29 May 2019.

p. 42 'We can't cross-examine in the dark': Michael McHugh SC for Crown, Supreme Court of the Northern Territory, trial of Margaret Pestorius and others, 21 November 2017.

TRESPASS

p. 43 Alice Springs Peace Group: 'Closed by the people', 19 October 1986 – 19 October 1987, Alice Springs Peace Group Bases Campaign, typescript held in the Alice Springs Public Library; Russell Goldflam, personal communication, email 24 March 2019.

Women's Peace Camp: Megg Kelham, 'War and peace: A case of global need, national unity and local dissent?: A closer look at Australia's Greenham Common', *Lilith: A Feminist History Journal*, vol. 19, 2013, p. 76.

p. 44 Pine Gap/Alice Springs a possible nuclear target: see Desmond Ball, *A Suitable Piece of Real Estate: American installations in Australia*, Hale & Iremonger, Sydney, 1980, ch. 12, 'Australia as a nuclear target'; see also Paul Dibb, *Inside the Wilderness of Mirrors: Australia and the threat from the Soviet Union in the Cold War and Russia Today*, Melbourne University Publishing, 2018, pp. 61–71. Australian governments 'remiss' on civil defence: Dibb, ibid., p. 151; see also Tom Gilling, *Project Rainfall: The secret history of Pine Gap*, Allen & Unwin, Crows Nest, NSW, 2019, ch. 18, 'Apocalypse Now'.

p. 45 Convergence, 2002: Australian Anti-Bases Campaign Coalition, 'Desert peace protest – October 5–7, 2002', Campaign Index, accessed online 16 May 2018; 'Missile Test Condemned', AABCC press release, 15 July 2001, accessed online 30 August 2018; 'Pine Gap Protest', AABCC press release, 20 September 2002, accessed online 28 August 2018.

p. 46 Campaign organisers in Alice Springs: Erwin Chlanda, 'Pine Gap in a troubled world', *Alice Springs News*, 11 September 2002.

p. 48 Arrests, melee: Erwin Chlanda, 'Huge media coverage of Pine Gap protest', *Alice Springs News*, 9 October 2002.

p. 51 'Bloody awful' journey: Bryan Law, Electronic Record of Interview (with police), 9 December 2005, personal archive of Margaret Pestorius.

Rocky Tiger Ploughshares action: Bryan Law, 'From Pine Gap to Rockhampton, via Waihopai: Actions, Verdicts, Lessons', and '5 Jane St, Depot Hill Q 4700', in his blog *Peace by Peace: We aren't gunna study war no more*, accessed online 30 August 2018.

NVDA principles: Bryan Law, 'Pine Gap – the Empire Strikes Back', *Webdiary*, 19 February 2007, accessed online 30 August 2018.

p. 52 Berrigan brothers and Ploughshares/Plowshares movement: '"Turbulent priest" fought for peace', *Sydney Morning Herald*, 10 December 2002; Zak, op.

cit., ch. 3, 'The prophet', ch. 4, 'The miracle'; Dan Berger, *The Struggle Within: Prisons, political prisoners, and mass movements in the United States*, Kersplebedeb Publishing and Distribution, Montreal, Canada, 2014, pp. 63–4; Sharon Erickson Nepstad, *Religion and War Resistance*, Cambridge University Press, New York, 2008, pp. 16–18; Frida Berrigan, 'How do you tell the kids that Grandma is in jail for resisting nuclear weapons?', *Waging Nonviolence*, 6 April 2018, accessed online 11 April 2018; Frida Berrigan, '50 years later, the spirit of the Catonsville Nine lives on', *Waging Nonviolence*, 16 May 2018, accessed online 17 May 2018; Harry J Cargas, 'Creating a community of risk: An interview with Elizabeth McAlister', *Commonweal Magazine*, 15 October 1971, accessed online 1 September 2018; William Plummer, 'Separated from her children, jailed nuke protester Liz McAlister says she's serving time for peace', *People*, 27 August 1984, accessed online 1 September 2018.

Ciaron O'Reilly: Joshua Robertson, 'Anti-war activist Ciaron O'Reilly: conventional protests are "a dead end"', *Guardian*, 7 January 2016, accessed online 5 September 2018; Kieran Finnane, 'Peace activist: have hammer, will travel', *Alice Springs News*, 12 October 2006; 'Aussie in court over US jet damage', *Sydney Morning Herald*, 27 May 2003, accessed online 5 September 2018; 'Peace activist slams ASIO treatment', *Sydney Morning Herald*, 8 February 2006, accessed online 31 August 2018; 'Pitstop Ploughshares finally acquitted', *Peace News*, issue 2477, September 2006, accessed online 3 September 2018.

p. 53 Kings Bay Plowshares: Mike W in On The Line, 'Kings Bay Plowshares indicted in Southern District of Georgia Federal Court', *LA Catholic Worker*, 4 May 2018, accessed online 31 August 2018; see also kingsbayplowshares7.org. For their risk of serious prison time: see Erickson Nepstad, op. cit., comparative table of Plowshares sentences by country, p. 18. On 28 October 2019 they were convicted by a Georgia federal jury of conspiracy, destruction of property on a naval station, depredation of government property and trespass. They were awaiting sentence at the time of writing: see Marjorie Cohn, 'Convicted anti-nuclear activists speak out: "Pentagon has brainwashed people"', *Truthout*, 28 October 2019, accessed online 30 October 2019.

Transform Now Plowshares: theirs is the central story in Zak, op. cit.

p. 54 Inspirational NVDA, on becoming Catholic, intentions for peace work in Rockhampton: Law, *Peace by Peace*, op. cit.

Medical conditions, letter to Hill, 2005 trespass: Law, Electronic Record of Interview, op. cit.

p. 55 'It was a miracle!': Donna Mulhearn, *The Pilgrim*, 10 December 2005 entry, accessed online 3 September 2018.

Jim and Adele's trespass in 2005: Jim Dowling, 'The bush track to Pine Gap', in Mulhearn, *The Pilgrim*, ibid., 9 December 2005 entry; account also archived on the Nautilus Institute site.

p. 57 Michael McKinley on Pine Gap's contribution to Iraq war: quoted in 'Christian pacifists to inspect Pine Gap', media release, 8 December 2005, in Mulhearn, *The Pilgrim*, ibid.

Human shield in Iraq: Mulhearn, 10 March 2003 message to supporters archived in *The Pilgrim*, ibid. 'The problem of achieving peace', Gandhi quote, inter-faith gathering: 17 March 2004 message to supporters, in *The Pilgrim*, ibid. 'Futility of violence': 15 May 2004 message to supporters, in *The Pilgrim*, ibid. See also Mulhearn's memoir: *Ordinary Courage: My journey to Baghdad as a human shield*, Pier 9, Millers Point, NSW, 2010.

p. 59 Law, Electronic Record of Interview, op. cit.

LAMENT

The narrative of the Peace Pilgrims' actions during the 2016 convergence, including their trespass and lament at the base and the police response to it, has been compiled from my interviews with them on multiple dates, their contemporaneous written accounts published on the Close Pine Gap webpage, and testimony (the police's and the Pilgrims's) from their trials. 'Peace Pilgrims – a Pine Gap tour diary' by Andy Paine has also been an important source, published on his blog andypaine.com on 23 August 2017, accessed 5 May 2018.

p. 63 Graeme Dunstan's tribute to Bryan Law, 'Death of a Peacemaker', and his biographic details on peacebus.com, accessed 9 May 2018.

p. 64 The poem that inspired the Pilgrims: Rainer Maria Rilke, 'The Tenth Elegy', *Duino Elegies*, translated by A. S. Kline. Excerpts reproduced with permission of Poetry in Translation, a not-for-profit organisation with the educational goal of increasing access to the classics. The permission is not intended as an endorsement of the actions of the protesters nor of this book.

Live-firing: see, for example, 'Talisman Saber [sic] 17 exercises Allies' capabilities' on Australian Army website, accessed 6 July 2019: 'TS17 will incorporate force preparation activities, Special Forces activities, amphibious landings, parachuting, land force manoeuvre, urban operations, air operations, maritime operations and the coordinated firing of live ammunition and explosive ordnance from small arms, artillery, naval vessels and aircraft.' However, there would be no live-firing in 2019, only blank and simulated ammunition would be used, according to Department of Defence TS19 webpage, Frequently Asked Questions, accessed 6 July 2019. About 34,000 troops from the US, Australia, Japan and eighteen other observer countries participated in TS19: see Paul Maley, 'Red, might and Blue to keep it real at Talisman Sabre', *Weekend Australian*, 13 July 2019, accessed online same date.

Simon Moyle quoting Merton in *Living a Costly Grace in the Face of War*, a film by Julian Masters, Vimeo, 5 February 2013, accessed 18 June 2018.

p. 65 Kevin Rudd, 'Faith in politics', *The Monthly*, October 2006.

p. 67 Disarm Collective, 'Close the eyes and ears of the US war machine in Central Australia', disarm.net.au, accessed 6 July 2019.

p. 75 Rilke, 'The Tenth Elegy', op. cit.

Napier is fit: Nick Kossatch, 'Kiwi Warriors notch double with 2015 Alice Springs rugby premiership', *NT News*, 30 March 2015, accessed online 17 April 2018; Alice Springs Cycling Club page on Flow Mountain Bike website, accessed 17 April 2018.

THE LAW

p. 77 On Mumu Mike Williams: Kieran Finnane, 'Stance of the patriarchs and

matriarchs', *Alice Springs News*, 19 April 2016. See also his memoir, *Kulinmaya! Keep listening, everybody!*, English translation by Linda Rive, Allen & Unwin, Crows Nest, NSW, 2019.
Radioactive cloud dispersed over entire mainland: Elizabeth Tynan, *Atomic Thunder: The Maralinga story*, NewSouth Publishing, Sydney, 2016, pp. 70, 87.
Radioactive legacy in our brains: Tim Flannery, *Here on Earth: An argument for hope*, Text Publishing, Melbourne, 2010, p. 158.

p. 78 'Not likely to be entirely free from contamination': British PM Clement Atlee in a message to Menzies on 26 March 1951, quoted in P N Grabosky, *Wayward governance: Illegality and its control in the public sector*, Australian Institute of Criminology, Canberra, 1989, ch. 16, 'A toxic legacy: British nuclear weapons testing in Australia', accessed online 31 July 2018.
Gross underestimate: For a detailed account of the devastation caused to Australian lands and people found by the Royal Commission and later investigations, see Tynan, op. cit.
McBride's speech: *Defence (Special Undertakings) Bill 1952*, Second Reading, Minister for Defence Philip McBride, House of Representatives Official Hansard, No. 23 1952, 4 June 1952, pp 1374–6, accessed online 30 July 2018.
Australia's interests assumed to coincide with Britain's: contemporaneous editorial, *West Australian*, quoted by Tynan, op. cit., p. 73.

p. 79 First atomic test: ibid., pp 70–1.
Montebello Marine Park 'explosive attraction' and safety advice: WA Parks and Wildlife Service website, accessed 8 July 2019. See also the islands' listing on dark-tourism.com, accessed 8 July 2019.
Islands not cleaned up and salvagers unprotected: Tynan, op. cit., p. 89.
'Rum Jungle Declared Prohibited Area', *Sydney Morning Herald*, 29 August 1952, p. 1, Accessed on Trove, 10 August 2018.
Several lots pieced together, they included one portion of unalienated Crown land: land title history provided on my request by Department of Infrastructure, Planning and Logistics, Northern Territory Government, 24 April 2018.
Pine Gap's early history: Tanter, 'Our poisoned heart: the transformation of Pine Gap', op. cit.; Tanter, 'The "Joint Facilities" revisited', op. cit. Secrecy around Pine Gap's purpose, early years: Ball, *A suitable piece of real estate*, op. cit., pp. 20–1. CIA's role revealed, recapped in Brian Toohey, *Secret: The making of Australia's security state*, Melbourne University Press, Carlton, Vic, 2019, ch. 19, 'The man who thought he owned the secrets' and ch. 20, 'The man who thought he owned a prime minister'; see also Tom Gilling, *Project Rainfall*, Allen & Unwin, 2019, Crows Nest, NSW, ch. 16, 'Never again'.

p. 80 Use of the *DSU Act* threatened before: Mary Heath, College of Business, Government and Law, Flinders University, Australia, provides evidence that activists were aware of the Act and its heavy penalties as far back as 1974, in an action at North West Cape in WA. She also gives the numbers for mass arrests during protests at Pine Gap in the 1980s, when trespass charges were laid under the *Commonwealth Crimes Act 1914*, as was the case at Nurrungar. See Mary Heath, 'Continuing the Cold War tradition and suppressing contemporary dissent', *Alternative Law Journal*, vol. 42, no. 4, 2017, pp. 248–52 at 250–1. The Alice Springs Peace Group in the '80s were also very aware of the legislation, writing about its infringement on local civil liberties, its use to stifle legitimate protest and media coverage of such protest and to allow surveillance of residents and activists: see 'Closed by the people', typescript, op. cit.

'Clandestine' actions, 'striking at the heart of national security': This was how the Crown characterised the actions of the Christians Against All Terrorism during their trial, according to Bryan Law's summary of the case dated 2007, personal archive of Margaret Pestorius. The second descriptor would be used again in the trials of the Pine Gap Peace Pilgrims.

'Blurred the separation of political and judicial powers': defendants quoted in Kieran Finnane, '"Security threat" if Minister thinks so', *Alice Springs News*, 19 October 2006. Further comment, Jim Dowling, personal communication, email, 9 May 2019.

Former Attorney-General Philip Ruddock, reply to questions received by email, 9 July 2019.

Ron Merkel QC background: Leonie Wood, '"Courageous" judge retires', *Age*, 16 May 2006, accessed online 6 February 2019.

p. 81 Pro bono assistance from Rowena Orr: Bryan Law, 'Pine Gap - the Empire Strikes Back', op. cit. Dr Ben Saul, then with the University of NSW's Centre of Public Law, also provided a pro bono legal opinion of possible defences under the Act, copy in the personal archive of Margaret Pestorius.

'Both exhilarating and challenging': Russell Goldflam, 'Satellites, citizens and secrets: R v Law & Others', Nautilus Institute, APSNet Policy Forum, 1 September 2008, accessed online 22 November 2017. This article gives an excellent account of the trial, the appeal and the legal issues arising.

On correspondence with defence minister, consequences: Bryan Law, 'Pine Gap – the Empire Strikes Back', op. cit.

p. 82 Ruling against challenge to 'prohibited area' status: Thomas J, *R v Law & Ors* [2006] NTSC 84, Reasons for Judgment, delivered 12 October 2006, paragraphs 26–7. See also Kieran Finnane, 'Pine Gap not "prohibited"?', *Alice Springs News*, 5 October 2006; and Kieran Finnane, '"Security threat" if Minister thinks so', op. cit. For further reported commentary on these issues, see Erwin Chlanda, 'No jail terms for Pine Gap Four', *Alice Springs News*, 21 June 2007. For further analysis, see Goldflam, op. cit.

Application for discovery and denial: Thomas J, *R v Law & Ors* [2007] NTSC 26, Reasons for Judgment, delivered 18 April 2007.

p. 84 Legal issues and arguments: drawn from the overview in appeal from Thomas J: *R v Law & Ors* [2008] NTCCA 4, No. CA 8 of 2007, Martin (BR) CJ, Angel and Riley JJ, Reasons for Judgment, delivered 19 March 2008.

Jury reaction: Goldflam, op. cit.

p. 85 Arrest and imprisonment for non-payment of fines: *Pine Gap on Trial* blog, 'Appeal Day Two', 22 February 2008, and 'Donna's thoughts re: Darwin arrest', 12 February 2008, accessed 5 February 2019.

Appeal result reported: Kieran Finnane, 'Pine Gap show and tell?', *Alice Springs News*, 28 February 2008.

p. 86 'The Crown have in effect given a warning': Thomas J, *R v Law & Ors*, SC 20529991, 20529975, 20529995 and 20529976, Sentence, 15 June 2007, transcript of proceedings.

Civil liberties, civil disobedience: Richard Ackland, 'Another bundle of intrusions', *Sydney Morning Herald*, 4 January 2008, accessed online 6 February 2019; Frank Brennan, *Too Much Order with Too Little Law*, University of Queensland Press, St Lucia, Qld, 1983; Frank Brennan, 'Civil disobedience a democratic safeguard', *Eureka Street*, vol. 18, no. 5, 7 March 2008, accessed online 12 January 2019.

p. 87 Amendments to the *DSU Act*: *Defence Legislation (Miscellaneous Amendments) Bill 2008*, Second Reading, House of Representatives Hansard, 10 February 2009, accessed online 22 May 2018.

p. 88 Garrett's changed position: Nick Grimm, 'Peter Garrett back flips on Pine Gap', *PM*, ABC, 10 June 2004, accessed online 11 January 2019.

p. 89 1987 visit by Labor MPs: 'Labor MPs back call for public debate and inquiry into US bases in Australia', press release (on Parliament of Australia, House of Representatives letterhead), 16 October 1987, digital scan in personal archive of Russell Goldflam; Edmund Doogue, 'Lease at US base should not be renewed, say MPs', *Age*, 17 October 1987.

Kieran Finnane, 'Snowdon's new demand for Pine Gap answers as major protest is planned', *Alice Springs News*, 13 March 2002; Erwin Chlanda, 'Back to the future with Warren Snowdon', *Alice Springs News*, 7 May 2019.

p. 90 *Defence Legislation (Miscellaneous Amendments) Bill 2008*, Second Reading, The Senate Hansard, 11 March 2009, accessed online 8 February 2019.

Raytheon at Pine Gap: Desmond Ball, Bill Robinson, Richard Tanter and Philip Dorling, 'The Corporatisation of Pine Gap', NAPSNet Special Report, 25 June 2015, accessed online 25 June 2019; see also Rosenberg, op. cit., pp. 40–1.

Australia kept in the dark: 'Report 26: An Agreement to extend the period of operation of the Joint Defence Facility at Pine Gap', Joint Standing Committee on Treaties, October 1999, pp. 5–6.

ENDS AND MEANS

p. 92 Swan Island military base: collated references under 'Australian defence facilities: Swan Island Training Area', Nautilus Institute for Security and Sustainability website, accessed 24 July 2018.

p. 93 Deaths of SAS soldiers, abuse of peace activists: Barry Hill, 'The uses and abuses of humiliation: Rabindranath Tagore's management of defeat', first published in *Griffith Review: Enduring Legacies*, edition 48, 2016; republished in *Reason & Lovelessness: Essays, encounters, reviews 1980–2017*, Monash University Publishing, Clayton, Vic, 2018, pp. 432–45. Barry Hill, 'Unspeakable heroes', poem, *Australian*, 14 November 2014, reproduced in 'The uses and abuses of humiliation', op. cit.

p. 94 Swan Island Peace Convergence, sources collated under '2 October 2014 Swan Island incursion', Australian Defence Facilities Briefing Books, 21 October 2014, Nautilus Institute for Security and Sustainability website, accessed 24 July 2018; Louise Milligan, 'Swan Island protesters claim they were hooded and stripped at ASIS base; Critics say "dumb" to trespass in heightened security climate', *7.30*, ABC, broadcast 9 October 2014, accessed online 24 July 2018.

Inquiry into the circumstances surrounding the trespass incident at Swan Island on 2 October 2014, conducted by Lieutenant-Colonel MA Kelly, report dated 22 January 2015, p. 8, paragraph 22, p. 9, paragraphs 24 & 28, redacted report accessed online 24 July 2018. The report does not contain the statements from the four protesters attached to the original, nor a paraphrase of their accounts.

'They wrote to their supporters': 'The Swan Island 3 settle: Still a #ToxicSAS?', *Wage Peace, Disrupt War,* undated, accessed online 24 July 2018.

p. 96 For Jim's accounts of these experiences recalled by Franz, see Jim Dowling, 'Raytheon 2 found "Not Guilty"', *Workers Bush Telegraph*, 26 July 2008, accessed online 1 August 2018; 'Public nuisance told to pay fine or do time', *ABC News*, 6 June 2005, accessed online 1 August 2018; 'Are Queensland's police out of control?', *Crikey*, 6 September 2005, accessed online 1 August 2018; Letter from the Office of the Assistant Commissioner Misconduct, 22 May 2008, posted by Jim Dowling in 'Queensland Crime and Misconduct Commission reveals its "Double Blind" investigation', *Workers Bush Telegraph,* 2 June 2008, accessed online 1 August 2018.

p. 97 Federal Agent Davey interviews Margaret: Electronic Record of Interview, 29 September 2016, personal archive of Margaret Pestorius.

p. 99 'It needs perimeter patrol': Senator Scott Ludlam, *Defence Legislation (Miscellaneous Amendments) Bill 2008*, Second Reading speech, The Senate Hansard, 11 March 2009, pp. 1331–2, accessed online 8 February 2019.

p. 100 Federal Agent Davey interviews Paul Christie: Electronic Record of Interview, 3 October 2016. This account is drawn from my contemporaneous notes when it was played for the court during Paul's trial.

p. 106 Hasty drafting of *DSU Act*: Tynan, op. cit. p 83; Heath, op. cit., p. 250.

DSU Act 'sledgehammer to crack a walnut': Goldflam quoted in Kieran Finnane, 'Red faces for federal police bringing peace activists to court', *Alice Springs News*, 1 October 2016.

p. 108 War crimes: McCormack and Durham in Tanaka and Young (eds), op. cit., pp. 219, 226–7.

'This was extraordinary': Goldflam quoted in Kieran Finnane, 'Into the breach', *Saturday Paper*, 9 December 2017.

Former Attorney-General George Brandis declined to answer questions: this advice was provided by the Communications & Public Affairs Adviser for the Commission, 15 May 2019.

p. 109 Tanter background, personal communication, 18 December 2018.

Goldflam background: Russell Goldflam, 'Pine Gap and the Nobel prize the Oz government ignores', *Alice Springs News*, 16 October 2017; 'Despite protests Pine Gap bigger, more powerful than ever', *Alice Springs News*, 5 October 2016; personal communication, multiple dates.

THE SACRED

p. 111 First World War centenary commemorations: Nationally there were 1650 projects worth close to $18 million across all 150 federal electorates. Minister for Veterans' Affairs and Minister Assisting the Prime Minister for the Centenary of ANZAC, Senator Michael Ronaldson, 'Anzac centenary local grants support community commemoration', media release VA049, 26 May 2015, accessed online 10 August 2018.

Australia's compulsory military service for home defence: 'Universal Service Scheme, 1911–1929', Australian War Memorial article, last updated 21 June 2019, accessed online 9 July 2019.

p. 112 Conscription referenda of 1916–17: See Ken Inglis, *Sacred Places*, Melbourne University Publishing, Carlton, Vic, 2008, pp 114–17. Referenda results: 1916: 1,087,557 in favour and 1,160,033 against; 1917: 1,015,159 in favour and 1,181,747 against; see National Archives of Australia, 'Conscription referendums, 1916 and 1917 – Fact sheet 161', accessed online 9 July 2019.

p. 113 Temple of Peace, rarity: Inglis, op. cit., pp. 232–3.
Richard Ramo: Queensland Heritage Register entry for the Temple of Peace, drawing on the work of Judith McKay, 'Brisbane's Temple of Peace: war and myth-making', *Queensland History Journal*, vol. 20, no. 10, May 2009, pp. 457–69, accessed online 16 November 2018.
The narrative for the Toowong Cemetery action and legal proceedings, apart from contemporary media reporting where indicated, draws on Andy Paine's detailed account, 'Put away your sword – repentance, prophecy, arrests and media outrage', andypaine.com, 13 February 2018, accessed 5 May 2018.

p. 115 First reports: Cameron Atfield, 'Religious fanatics vandalise war memorial', *Brisbane Times,* 2 March 2017, accessed online 3 March 2018; 'Crusading cowards charged with smashing up a Brisbane War Memorial', *10 News First,* 2 March 2017, accessed online 22 November 2018.

p. 117 Jim saddened: Andrew West, 'Is it a peace action or a desecration?' *Religion and Ethics Report,* ABC Radio National, 12 July 2017, accessed online 3 March 2018. This interview was played during the trial. See p. 125.
Jonathan Sri on the Catholic Workers: *Ten's Eyewitness News*, 2 March 2017, viewed online 5 May 2018.

p. 118 For more on Graeme Dunstan's Frontier Wars remembrance see his blog: peacebus.com.

p. 119 The Sovereign Yidindji Government: Paul Daley, 'Murrumu: one man's mission to create a sovereign Indigenous country inside Australia', *Guardian*, 6 March 2016, accessed online 2 November 2018; Paul Daley, 'Indigenous activist Murrumu has fought the law this week. But who will win?', *Guardian*, 8 August 2015, accessed online 15 February 2019.
A Chorus of Women began when some 150 women filled the Australian Parliament on 18 March 2003, the day Australia's intention to invade Iraq was announced, and sang 'Lament' by Glenda Cloughley and Judith Clingan AM: see their website (https://www.chorusofwomen.org/), accessed 15 February 2019.

p. 120 Anzac Day Dawn Service, Cairns, 25 April 2016: video by Cairns Regional Council, *YouTube*, 5 May 2016, accessed 15 November 2018.
'Frontier Wars Ceremony – How we organised in Gimuy', video by Margaret Pestorius, *YouTube*, 22 March 2018, accessed 3 November 2018.

p. 121 Alice Springs Anzac memorial works: Kieran Finnane, 'Telling the stories of war: we could do so much better', *Alice Springs News*, 23 January 2017.
Local streets named after men known to have carried out killings of Aboriginal people, examples: Mounted Constable William Willshire, after whom a street is named: see DJ Mulvaney, 'Willshire, William Henry (1852–1925)', *Australian Dictionary of Biography*, Melbourne University Press, Carlton, Vic,

1990, vol. 12; Erwin Chlanda, 'Was Willshire a murderer?', interview series with RG (Dick) Kimber, *Alice Springs News*, 20 & 27 November, 4 December 2008. William Benstead, after whom a street, a creek and a nearby mountain are named: see Stuart Traynor, *Alice Springs: From singing wire to iconic outback town*, Wakefield Press, Mile End, SA, 2016, pp. 123–4; Sam Gason is another remembered in a street name; see Traynor, ibid., pp. 96–9.

Stuart statue saga: Kieran Finnane, 'Stuart statue: should council give it back?', *Alice Springs News*, 11 February 2014. The article includes links to the main earlier reports.

2019 Frontier Wars remembrance in Alice Springs: Kieran Finnane, 'Recognising the forgotten wars, necessary for peace', *Alice Springs News*, 25 April 2019.

p. 122 Speeches at the official Cairns Anzac Day ceremony: video by Cairns Regional Council, op. cit.

p. 124 Reports of the Christians' trial for the Toowong Cemetery action: Tony Moore, 'Christian tells court he had 'higher permission' to remove sword from Toowong memorial', *Brisbane Times*, 19 July 2017, accessed online 23 November 2018; Mark Bowling, 'Court trial continuing for Catholic men charged with wilful damage of Toowong war memorial cross', *Catholic Leader*, 20 July 2017, accessed online 3 March 2018; Tony Moore, '"You cannot hide your guilt behind your beliefs": Magistrate', *Brisbane Times*, 24 July 2017, accessed online 3 March 2018; Mark Bowling, 'Catholic Workers found guilty over damaging war memorial', *Catholic Leader*, 26 July 2017, accessed online 23 November 2018.

A RECKLESS ACT OF PRAYER

p. 129 Pre-trial hearings: Russell Goldflam, personal communication, email, 11 November 2017; inspection of his notes from the 3 November 2017 hearing.

p. 130 Court 'materials' on Justice John Reeves and Pine Gap: Scott Campbell-Smith, 'Pine Gap: to spy or not to spy', *Alice Springs News*, 27 March 2002; Erwin Chlanda, 'Aboriginal land council wants deals powers for tribal regions', *Alice Springs News*, 18 November 1998; Philip Nitschke with Peter Corris, *Damned If I Do*, Melbourne University Publishing, Carlton, Vic, 2013.

p. 132 Japanese preparing to surrender before atomic bombs were dropped: see, for example, Lindqvist, op. cit., pp. 110–12, sections 231–6, pp. 114–15, sections 241–2, pp. 117–18, sections 249–50.

p. 133 Ruby, 'Minding the Gap', op. cit., pp. 8–10, accessed online 1 March 2019. At the time of writing, the title for her dissertation was 'The Fourth Eye: Australia's Role in the Five Eyes before and after Snowden': see felicityruby.com.

Supreme Court building: Kieran Finnane, 'Court placing itself above the community?', *Alice Springs News*, 5 May 2017.

p. 136 Report excerpt, Kieran Finnane, 'Trial of Peace Pilgrim begins', *Alice Springs News*, 13 November 2017.

p. 139 Reader comments, ibid.

p. 140 Reader comments: Kieran Finnane, 'Peace Pilgrim says he had permission to be on Pine Gap', *Alice Springs News*, 14 November 2017.

p. 146 The Queen v Saleh: *R v Saleh* [2015] NSWCCA 299.
Fairford Five news reports: George Monbiot, 'Protesters who have damaged military equipment are walking away from the dock', *Guardian*, 17 October 2006, accessed online 7 June 2018; Fred Attewill, 'Court clears anti-war saboteurs', *Guardian*, 23 May 2007, accessed online 7 June 2018; Esther Addley and Richard Norton-Taylor, 'Fairford Two strike blow for antiwar protesters after jury decide they were acting to stop crime', *Guardian*, 26 May 2007, accessed online 7 June 2018; Richard Norton-Taylor, 'RAF base saboteurs lose court case', *Guardian*, 7 July 2007, accessed online 7 June 2018.

p. 147 Lord Hoffman on civil disobedience: Opinions of the Lords of Appeal for Judgment in the Cause, House of Lords, Session 2005–06 [2006] UKHL 16, accessed online 8 March 2019.

p. 150 Russell Goldflam on the possibility of a lower court hearing: personal communication, email, 1 December 2017.

p. 151 Reader comments: Kieran Finnane, 'Guilty: unanimous jury verdict for Peace Pilgrim', *Alice Springs News*, 15 November 2017.

NOTHING TO SEE HERE

p. 155 DVD copy of CCTV footage sent to Russell Goldflam: personal communication, 7 November 2018.

p. 161 Daniel O'Gorman SC available to advise them to the end: Goldflam, personal communication, 7 November 2018.

p. 163 Andrew Bartlett, Senate Adjournment Speech, 16 November 2017, Senate Hansard, p. 8752, accessed online 17 November 2017.
Open Letter to Attorney-General Brandis, *Saturday Paper*, 11 November 2017.

p. 165 Peter Wallace Peltharre, biographic note for the Parrtjima Festival: 'My Father's country is Alatyeye, Yam Country, Gem Tree. I'm Apmereke-artweye for that. My Mother's country is Antulye (Undoolya), Eagle Dreaming: that's my Mother and Grandmother. Kukatye, Gum Tree Dreaming, that's my Grandfather. I'm Kwertungurle for that. The sacred site near Pine Gap, that's my Mother's and Grandfather's. I've seen people standing there, like soldiers, you know, like that ... but that's my Mother and Grandfather's country', accessed online 5 March 2019.

p. 167 'Remember Ken?' Donna Mulhearn, *Pine Gap on Trial* blog, 'Appeal Day 1', 20 February 2008, accessed online 5 February 2019.
Jo Vallentine comment: Kieran Finnane, 'Strange encounters: the Peace Pilgrim and the Police Sergeant', *Alice Springs News*, 18 November 2017.

ORDINARY OR EXTRAORDINARY?

p. 186 Interview with Malcolm Fraser: *RN Breakfast*, ABC Radio National, 28 April 2014, and related article by Matt O'Neil, 'Malcolm Fraser calls for closure of Pine Gap Joint Defence Facility', accessed online 20 November 2017.

p. 193 Philip Dorling, 'Pine Gap drives US drone kills', *Sydney Morning Herald*, 21 July 2013. Jeremy Scahill on using metadata for drone strikes: see 'The Assassination Complex', *Intercept*, 15 October 2015. Brandon Bryant: see, for example, Ed Pilkington, 'Life as a drone operator: "Ever step on ants and never give it another thought?"', *Guardian*, 20 November 2015.

p. 194 Zubair: see, for example, Alexander Abad-Santos, 'This 13-year-old is scared when the sky is blue because of our drones', *Atlantic*, 29 October 2013.

p. 202 Robert Fisk, 'Amid allied jubilation, a child lies in agony, clothes soaked in blood', *Independent*, 8 April 2003, accessed online 12 July 2019.
Eric Margolis, 'Pentagon and CIA at war in Syria', 3 September 2016, archived on ericmargolis.com, accessed 12 July 2019.

p. 203 Land title history: More than one pastoral lease was involved but 325 acres were unalienated Crown land: Hansard, 23 April 1969, quoted by Gilling, op.cit., pp. 21–2.

p. 204 Kieran Finnane, personal communications with Peter 'Coco' Wallace Peltharre, 14 May & 14 October 2019.

p. 205 'Surveillance images of drone strikes in Pakistan', *Washington Post*, 26 April 2010, accessed online 13 October 2019.

CONSCIENCE VERSUS REASON

p. 212 Aerial bombing excluded from Nuremberg and Tokyo trials: Selden, op. cit., p. 92.

p. 213 Burgess and Saunders' 'No war' graffiti on Opera House: see, for example, Karen Michelmore, 'A subversive spotlight', 702 ABC Sydney, 4 February 2012, accessed online 17 April 2019. Removal of self-defence as a legal defence: *R v Burgess* (2005) 152 A Crim R 100, discussed by Thomas J in her judgment, *The Queen v Bryan Joseph Law & Ors* [2007] NTSC 45, paragraphs 21–4.

p. 215 Kieran Finnane, 'No extraordinary emergency at Pine Gap: judge rules', *Alice Springs News*, 22 November 2017.

p. 221 Kieran Finnane, 'Even if "unfair, unreasonable or too harsh", it is still the law', *Alice Springs News*, 23 November 2017.

p, 222 Comment by 'Jack': ibid.
Merciful verdicts: Russell Goldflam, personal communication, email, 30 April 2019.

p. 228 Ingo Müller 'has convincingly argued that members of the German judicial system were enthusiastic supporters of National Socialism', wrote Tom Munro in a review of Müller's book, *Hitler's Justice: The courts of the Third Reich* (translated by Deborah Lucas Schnieder, published by Harvard University Press, Cambridge, MA, 1994), *Alternative Law Journal*, vol. 48, 2000, accessed online 23 April 2019.

p. 229 Letter to the editor: Chris Hawke and Joy Taylor, 'Breaking the violence silence: at home, around the world', *Alice Springs News*, 29 November 2017.

p. 230 Curtis LeMay: Lindqvist, op. cit., p. 75, section 165, pp. 107–9, sections 223–8; see also obituary, 'Curtis LeMay, 83, Bomber General of WW II, dies: Warrior: He later built the Strategic Air Command. He was George Wallace's running mate in 1968', *Los Angeles Times*, 2 October 1990, accessed online 23 April 2019.

FIELDS OF ACTION

p. 231 Courtroom in Brisbane: Shae McDonald, 'Pine Gap trespassers are fined but avoid jail', *Brisbane Times*, 4 December 2017, accessed online 23 April 2019. On sentencing see also Kieran Finnane, 'Into the breach', *Saturday Paper*, 9 December 2017.

p. 232 Sentencing remarks: Reeves J, Transcript of Proceedings, at Alice Springs on Monday 4 December 2017, transcribed by DTI, published on the Supreme Court of the Northern Territory website, accessed 14 December 2017.

p. 233 Civil disobedience in Australia: Miriam Cosic, 'Offshore detention and the new face of civil disobedience', *Drum*, ABC, 8 February 2016, accessed online 26 April 2019; Sarah Whyte and Uma Patel, 'Doctors to launch High Court challenge against detention secrecy laws', *7.30*, ABC, 26 July 2016, transcript accessed online 27 April 2019; 'What are the secrecy provisions of the Border Force Act?', *ABC News*, 27 July 2016, accessed online 27 April 2019; Gareth Hutchens, 'Dutton retreats on offshore detention secrecy rules that threaten workers with jail', *Guardian*, 14 August 2017, accessed online 27 April 2019.

Medevac legislation: Katharine Murphy, 'Centre Alliance senator warns Coalition not to repeal medevac law', *Guardian*, 2 July 2019, accessed online 11 July 2019.

Franklin River blockade: see, for example, 'Franklin Dam and the Greens', Defining Moments, National Museum of Australia, article on their website, accessed 26 April 2019.

Draft resistance: see, for example, the Australian Living Peace Museum, multiple articles.

p. 236 Des Ball's last public interview: Hamish McDonald, 'Mind the Gap', *Saturday Paper*, 1–7 October 2016.

Hosting Pine Gap an all or nothing situation: Tanter, personal communication, 18 December 2018, and by email, 28 May 2019; see also Richard Tanter, 'An Australian pathway through Pine Gap to the nuclear ban treaty', *Pearls & Irritations*, 5 August 2019; *Alice Springs News*, 6 August 2019 (an extended and footnoted version is on the Nautilus Institute site).

p. 237 Treaty signatories and ratifications as at 24 March 2020, see treaties.un.org.

p. 239 Reasons for rejection of his Blue Card renewal: Franz Dowling, personal communication, 29 November 2019.

'Irrelevant conviction': Russell Goldflam, personal communication, 31 January 2020.

p. 240 '35 Years of Food Not Bombs', video on their website and historical timeline, accessed 30 October 2018.

EPILOGUE

p. 250 Margaret MacMillan, 'The Mark of Cain', The Reith Lectures 2018, broadcast and podcast by BBC Radio 4.

James Hillman, *A Terrible Love of War*, Penguin Books, New York, 2004, especially pp. 15–16, 30–1, 86–7.

p. 251 Jane Addams: see Debra Michals' entry for Addams on the website of the National Women's History Museum (in Alexandria, Virginia, USA), accessed 25 April 2019.

Jacqueline Chlanda, 'Nothing comes into the universe / and nothing leaves it: Maternal Subjectivity in the Poetry of Sharon Olds and the Art of Janine Antoni', PhD thesis, University of Queensland, 2019, unpublished.

p. 252 'About Woomera Prohibited Area': Australian Government, Department of

Defence, Woomera Prohibited Area Coordination Office, website accessed 14 July 2018.

p. 253 Militarised swathe: City of Salisbury, 'A year of economic change', accessed on their website, 5 April 2019; Andrew Green and Michael Park, 'Secret plans for new port outside Darwin to accommodate visiting US Marines', *ABC News*, 23 June 2019, accessed online 11 July 2019; Northern Territory Government, 'Defence and defence support', The Territory website, accessed 11 July 2019; Kieran Finnane, 'Defence industry to Alice: get a slice of the action', *Alice Springs News*, 21 November 2019.

p. 254 One example of weapons testing: Matthew Grimson and Mark Corcoran, 'Taranis drone: Britain's $336m supersonic unmanned aircraft launched over Woomera', *ABC News*, 7 February 2014.

Kristian Laemmle-Ruff's exhibition *Woomera* was shown at the Araluen Arts Centre in Alice Springs, 13 July – 19 August 2018. For a slightly edited version of my opening speech see Kieran Finnane, 'Artist's Woomera warning: live bombs', *Alice Springs News*, 14 July 2018.

ACKNOWLEDGEMENTS

As this book would not exist without him, Erwin Chlanda deserves to come first. A constant model of forthright and persistent journalistic endeavour, he has also been my 'funding body' and has carried the *Alice Springs News* with a much reduced contribution from me. He remains my ever loving daily companion.

My special gratitude goes to the following:

Barry Hill who recognised the book in the Pilgrims' story from the start. A generous and loyal friend of decades, he read and commented on the manuscript for the two years of its writing, always ready to demand more in understanding and nuance, always encouraging.

Tom Griffiths and Richard Tanter who wrote eloquent letters of support for my proposal to UQP. Richard remained ever ready to draw on his expertise in many of the subject areas this book delves into, pointing me in the right direction for further research and giving precise feedback on the manuscript.

Russell Goldflam, a longtime friend, and entwined in the Pilgrims' story as well in the history of Pine Gap protest and its prosecution. He read much of the manuscript with his habitual rigour, was incisive in his critique and generous in response to my every enquiry.

My brother Mark Finnane who offered important early guidance on the writing as well as challenges on perspective. He and my sisters,

Antonia Finnane, Gabrielle Finnane, Francine Finnane and Rebecca Finnane, have helped sustain me, in this project as throughout my life, with their loving interest and support.

My dear friend Jennifer Taylor who read the draft manuscript from start to end at a critically important time. She took part in the 1983 Women's Peace Camp at Pine Gap and was one of the 111 'Karen Silkwoods' arrested for trespass. I don't deal with these events but they are part of what provided a deep grounding for her insightful comments on the narrative. On a memorable Sunday afternoon, she also organised a reading, with her partner Sue Fielding and friend Annie Bolitho, of a troublesome chapter. Its improvement owes much to their feedback.

My dear friend Fiona Walsh who memorably walked and talked with me at Kweyernpe, among many wonderful walks and conversations. She has taught me much about this place we live in.

Peter Coco Wallace Peltharre, kwertengerle for the sacred area near Pine Gap, who generously answered my questions, asked his own and offered insightful reflections on the issues.

For making this a better book and encouraging me through its writing, I must also mention in alphabetical order: Cate Adams, Ruth Apelt, Alexandra Barwick, Gail Chlanda, Janet Chlanda, Julie Chlanda, Laurie Chlanda, Lesley Chlanda, Graeme Dunstan, Maureen Finnane, John Fitzgerald, Maria Giacon, Mike Gillam, John Hughes, Megg Kelham, Alex Kelly, Dick Kimber, Alex Nelson, Kerrie Nelson, Pip McManus, Kim Mahood, Rod Moss, Dani Powell, Anne Rampa, Libby Robin, Craig San Roque, Suzanne Smith.

For their friendly professional assistance: Emily Howie, Human Rights Law Centre; Xavier La Canna, NT Department of Attorney-General and Justice; Elisabeth Marnie, NT Archives Service; Marie Rançon, Central Land Council; Ben Taylor, NT Department of Infrastructure, Planning and Logistics; Alice Woods, Alice Springs Public Library; and the law courts staff in Alice Springs.

For their ongoing interest in my work: Ashley Hay and *Griffith Review*, in particular for publishing 'Reading in the dark', which helped shape parts of my chapter 'In the dark'. Also *The Saturday*

Paper for their publication of 'Into the breach', a feature article about the prosecutions of the Peace Pilgrims, which helped me gauge the broader interest in these events.

To the team at UQP, my thanks for backing me as a writer for a second time. Non-fiction publisher Alexandra Payne responded with enthusiasm to the book proposal, took it through to acquisition, and encouraged me through the early stages of writing. After she left UQP, I passed into the capable hands of publishing director Madonna Duffy, who kept me focussed on the book's goals. Managing editor Jacqueline Blanchard helped me see the forest through the trees of the manuscript and stayed the course with warm support through to the end. It was also a pleasure to work with UQP's commissioned copy editor, Nikki Lusk, for her interest in and grasp of the whole project, and the careful advice that followed.

To the Peace Pilgrims, Margaret Pestorius, Jim Dowling, Andy Paine, Franz Dowling, Tim Webb and Paul Christie, my deepest thanks for welcoming me into their lives and for their support and interest throughout the process of this book. It has been an inspiration and a delight to write their story.

To Kristian Laemmle-Ruff, my deep appreciation for his awakening photograph and his generosity in reserving it for the cover of this book.

To Anna Krien, my warmest thanks for her attentive reading of the manuscript as the COVID-19 crisis engulfed our lives and for understanding its perspective and relevance.

Finally, to my daughter Jacqueline Chlanda and son Rainer Chlanda. They have been with me every step of the way through the writing, as in all parts of my life, asking their searching questions, offering their perspectives, their information and insights, and above all their love. They are at the centre of my hope for peace in the world – the nurture of life.

TROUBLE
Kieran Finnane

What is going on in the often troubled town of Alice Springs?

On the streets of Alice Springs, in town camps, drinking camps and out on the highway, in gatherings awash with alcohol, men kill one another in seemingly senseless acts of aggression and revenge. Men kill their wives, families feud, women join the fighting and, in the wings, children watch and learn. From the ordered environment of the courtroom, *Trouble* lays out in detail some of this deep disorder in the town's recent history.

Drawing on her decades as a journalist in Central Australia, as well as experience of its everyday life, Kieran Finnane recognises a story beyond the horror and tragedy of the events, the guilt or innocence of perpetrators, to witness a town and region being painfully remade. In this groundbreaking book, we hear the voices from Australia's troubled heart and gain a unique insight into the challenges and potential of this hard and beautiful place.

'Finnane is precisely the type of journalist Australia needs but who is often overlooked by the metropolitan elites. She is enmeshed in the news as it breaks around her in a community of 30,000 people.'—*The Saturday Paper*

'This is journalism of the highest calibre.'—*Australian Book Review*

'... grimly fascinating and deeply thoughtful ...'—*The Australian*

ISBN 978 0 7022 5403 1